AUVSI XPONENTIAL 2022

Orlando, Florida, USA
25 – 28 April 2022

ISBN: 978-1-7138-5223-0

Printed from e-media with permission by:

Curran Associates, Inc.
57 Morehouse Lane
Red Hook, NY 12571

Some format issues inherent in the e-media version may also appear in this print version.

Association for Unmanned Vehicle Systems International
2700 South Quincy Street
Suite 400
Arlington, VA 22206
USA

Phone: (703) 845-9671
Fax: (703) 845-9679

www.auvsi.org

Additional copies of this publication are available from:

Curran Associates, Inc.
57 Morehouse Lane
Red Hook, NY 12571 USA
Phone: 845-758-0400
Fax: 845-758-2633
Email: curran@proceedings.com
Web: www.proceedings.com

TABLE OF CONTENTS

AERPAW – AERIAL EXPERIMENTATION AND RESEARCH PLATFORM FOR ADVANCED WIRELESS

Thomas Zajkowski[1]

AERPAW is a $24 million grant awarded by the PAWR Project Office on behalf of the National Science Foundation to develop an advanced wireless research platform, led by North Carolina State University, in partnership with Wireless Research Center of North Carolina, Mississippi State University, and Renaissance Computing Institute (RENCI) at the University of North Carolina at Chapel Hill; additional partners include Town of Cary, City of Raleigh, North Carolina Department of Transportation, Purdue University, University of South Carolina, and many other academic, industry and municipal partners. On the AERPAW platform, drones and 5G are integrated to be mutually beneficial. Drones support 5G by providing increased coverage and connectivity, and 5G supports drones by providing improved signals and location data.

INTRODUCTION

The Aerial Experimentation and Research Platform for Advanced Wireless (AERPAW) will support advanced and emerging wireless research for both aerial and ground vehicles. Academic research on unmanned aerial systems (UAS) communications has been increasing over the past decade to meet the expected growth of UAS for diverse applications. For example, according to IEEE's database, journal and conference publications in UAS communication grew an average of 29 per year from 1992 to 2009 to over 480 per year from 2015 to 2019, all with trends pointing to exponential growth in the future. Major telecommunications conferences and journals are now organizing special sessions, workshops, and special issues dedicated to UAS communications and networking. The IEEE Vehicular Technology Society (VTS) formed an ad-hoc committee on UAS in 2018 to coordinate research activities in this area. The 3GPP standardization recently studied the suitability of LTE networks to support aerial wireless devices currently looking for the integration of UAS into 5G.

To help assess the needs and requirements of potential users, we surveyed 71 researchers from academia, industry, and industrial labs. Survey results informed the design of AERPAW and indicated the types of experiments we must support. In terms of research opportunities, three topics received the highest interest 1) UAS integration into cellular and 2) vehicular infrastructure, and 3) wireless localization, identification, and tracking of and with UAS. There was also significant interest in using software-defined radios (SDRs), both at cell towers and, more critically, for the UAS. UAS applications such as search and rescue, detection, tracking, and use as a hot-spot base station (BS) received high-interest responses. Additional identified research topics of interest to global researchers included the following: mobile virtual reality/augmented reality (VR/AR); use

[1] Project Coordinator, Institute for Transportation Research and Education, North Carolina State University, Centennial Campus, Box 8601, Raleigh, NC 27695-8601

of wireless communications for control and guidance of multi-agent networking; energy endurance of aerial platforms; reflection measurements from ground and clutter; wireless energy transfer from UAS to sensors; multi-band connectivity with UAS; high-resolution imaging and surveying; localization and communication; computational and energy-efficient UAS protocols; dynamic and adaptive bandwidth allocation for UAS search and rescue operations; design of multi-band multi-function radio-frequency (RF), microwave, and millimeter-wave (mmWave) components and circuits; and extended UAS command and control using cellular and wireless local-area networks (WLAN).

In terms of experimental needs, the survey indicates the following key desired features for a testbed: 1) access to training materials, 2) the ability to reserve the testbed, 3) the ability to access the platform through multiple means (web, ssh, etc.), 4) remote access to SDRs, and 5) access to mmWave SDRs and fifth-generation (5G) new radio (NR) mobiles. These are taken into account in AERPAW's design. Figure 1 presents different experimentation possibilities intended to be supported by AERPAW once it is fully operational.

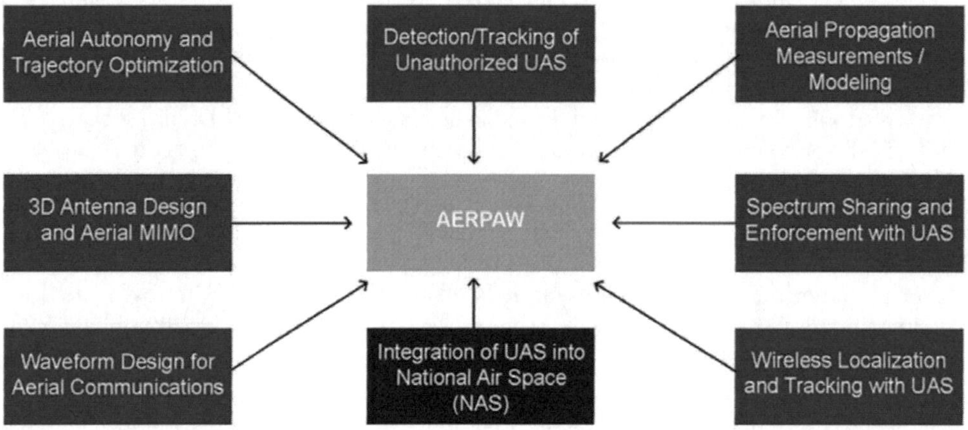

Figure 1. Overview of research experiments to be enabled by AERPAW

SCIENTIFIC REQUIREMENTS

AERPAW is envisioned as a platform for experimentation in support of advanced wireless experiments, as discussed in the previous section, to promptly deliver high-quality, accurate data to experimenters. To support such objectives and enable experiments at scale in a mobile environment with UAS and ground vehicles, various requirements and considerations should be considered, as summarized in Figure 2.

To enable advanced wireless experiments, AERPAW will provide access to advanced radios of various types, together with other hardware and software resources required to support sophisticated experimentation, including drivers, software development kits (SDKs), experiment templates, compute and storage resources, connectivity, and a variety of real-world spatial characteristics. Through user and technical literature surveys and existing partnerships, we have identified a broad range of specific experiment scenarios as described in the previous section and derived standard testbed requirements and the technical architecture to serve these. A unique feature of AERPAW is the support of mobile radio endpoints; both ground and airborne are considered a primary design determinant. Wireless communication is primarily driven by the need to be "untethered," and mobility is the root driver of many of the most difficult intellectual challenges in wireless theory and technology, from coarse-grained concerns such as handover to fine-grained ones such as beam-steering to track mobile nodes. AERPAW is unique in its combination of being embedded in natural outdoor spaces, various radio terrain ranging from agricultural to urban, and native support of wireless mobility in experiments. Further aspects of the technical architecture are driven by the paramount need for operational safety of airborne mobile nodes, the reproducibility of experiments, and the requirement to onboard users easily onto this unique environment on an ongoing basis.

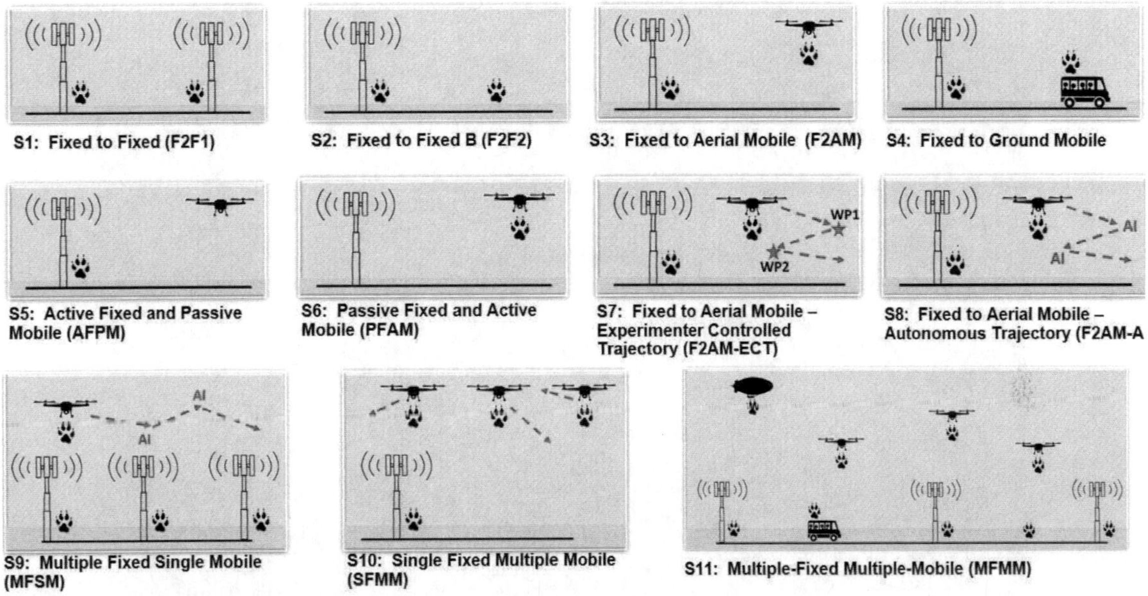

Figure 2. Overview of research experiments to be enabled by AERPAW

FACILITY AND INFRASTRUCTURE

AERPAW is a multi-institutional team: North Carolina State University (NC State is the lead university), Mississippi State University (MSU), the Wireless Research Center of North Carolina (WRC), RENCI, the Town of Cary, the City of Raleigh, Purdue University, and the University of South Carolina (USC). AERPAW is headquartered at NC State.

Research Triangle Park AERPAW Infrastructure

In the Research Triangle region of North Carolina, the AERPAW sites are as follows: 1) NC State University Centennial Campus (a suburban/campus site) and 2) Lake Wheeler Field Laboratories (located south of Centennial Campus, a rural and agricultural site). Based on the feedback from NSF and the PPO, the construction of towers in the Lake Wheeler area has been prioritized in Phase-2 to prioritize drone experiments that are more difficult to carry out in the Centennial Campus in the short term Figure 3.

MSU Development Infrastructure

MSU will develop and test technology for the AERPAW testbed at NC State. Therefore, infrastructure deployment at MSU is limited and aims to support the development and testing. We propose to use the same technology and systems as proposed for the AERPAW testbed, specifically SDR, 5G, and IoT systems, to facilitate the development of experimental tools. This requires hardware-software codesign for portability and interoperability and the assembly of AERPAW radio development prototypes.

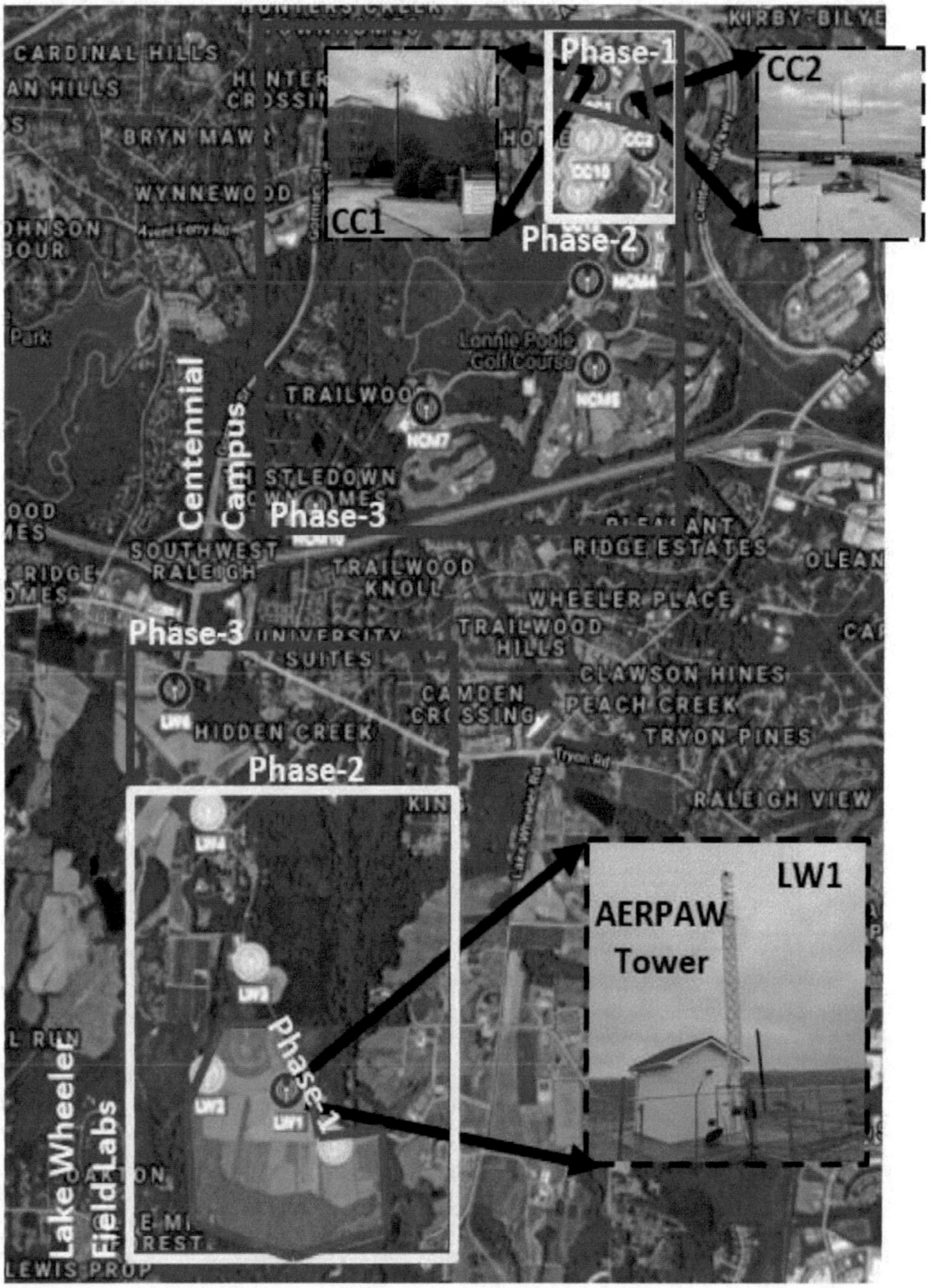

Figure 3. AERPAW Fixed Node Locations

DESIGN AND DEVELOPMENT

AEPAW execution will involve the platform's design, development, deployment, and operation. An overview of the envisioned platform is provided in Figure 3, which illustrates a typical AERPAW fixed tower, a mobile node that can be an aerial or ground vehicle with one or more radios, a control center, AERPAW backplane, data store, and the sandbox. Either the fixed tower and/or the mobile nodes can have one or more SDR equipment, 5G base station, RF sensor, IoT access station, etc.

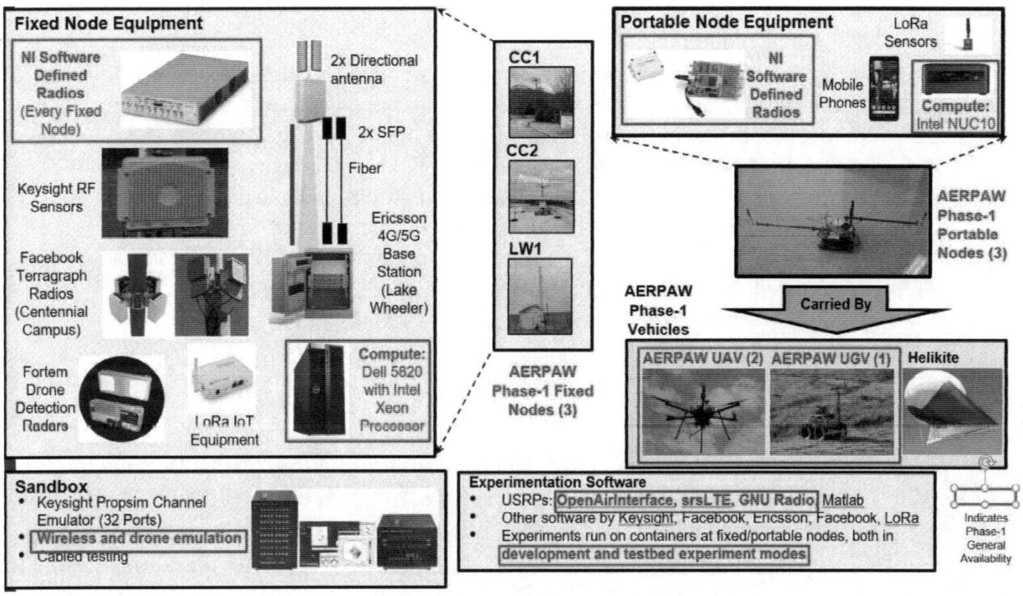

Figure 4. AERPAW Equipment

CONCLUSION

This paper presents a brief high-level overview of the AERPAW platform. The AERPAW platform will be able to support experiments where drones and advanced wireless technologies such as 5G are integrated to be mutually beneficial in a wide variety of applications. Drones support 5G by providing increased coverage and connectivity, and 5G supports drones by providing improved signals and location data. As an example, AERPAW is supporting National Radio Dynamic Zone (NRDZ) research.

REFERENCES

AERPAW Main Website: https://aerpaw.org

CHALLENGES AND PROGRESS IN SWARM AUTONOMY

David Koelle*

The growing use of uncrewed and increasingly autonomous vehicles across civilian and military applications starts to establish a future in which multiple vehicles can work together to achieve tasks. To bring these multi-robot teams, or swarms, into physical existence to aid human workers, several challenges must be addressed, and are currently under active research and development in academia and industry. These include notions of heterogeneity in the composition of swarms, task allocation, and challenges with real-world communications. This manuscript surveys the considerations and challenges in these areas and presents information about work that Charles River Analytics has performed on swarm systems through our Swarm Coordination Framework capability.

INTRODUCTION

The growing use of uncrewed and increasingly autonomous vehicles across civilian and military applications starts to establish a future in which multiple vehicles can work together to achieve tasks. To bring these multi-robot teams, or swarms, into physical existence to aid human workers, several challenges must be addressed, and are currently under active research and development in academia and industry. Currently, we have seen swarms as pre-orchestrated, choreographed displays of single-purpose vehicles that are increasingly popular at large celebrations (e.g., Shanghai's 2020 New Year's celebration) and major sporting events (e.g., Super Bowl LII in 2019). To understand the future potential of swarms, we survey the progress being made on work that will allow multi-robot teams to grow beyond choreographed displays through task allocation and centralized or decentralized organization. We will see how task allocation can be applied to swarms that consist of vehicles with different capabilities that change in health and status during the span of an operation. We will see how challenges in communication can be addressed, and how human operators will use swarms effectively through human-swarm interaction.

The increase in potential applications of multi-robots systems is driven by several factors, particularly the decreasing cost and increasing agility of uncrewed aerial vehicles (UAVs) and the increasing capabilities of these vehicles in terms of computational power. Sensors that can be placed on UAVs are also becoming increasingly capable, which includes advances in processing live video. These capabilities open a path for these vehicles to act with increasing degrees of autonomy, however there are two barriers to autonomy today: the state of autonomy itself, and policies put in place to protect people and property from errant vehicles. Advances in autonomy are underway across vehicles of all types (with industry putting substantial backing behind self-driving cars) and has been a goal of researchers (and a dream for the imaginative) for a several generations. Within the past couple of decades, the availability of vehicles that can realistically perform autonomous operation has helped get ideas in autonomy out from academic laboratories and into physical

---

* Director of Engineering, Charles River Analytics, 625 Mount Auburn St, Cambridge MA 02138

environments, where challenges become increasingly difficult. Full autonomy is not yet a solved problem, and regulations reflect this: In the United States, FAA Part 107 requires that an operator always maintains direct control of a UAV, and that the vehicle operates within visual distance of the operator. While one can request an exemption, confidence in autonomous operation must be significantly increased before policies allow autonomous operation without exemptions.

THE COMPOSITION OF SWARMS

How many vehicles need to be in a swarm to call the group a "swarm" in the first place? In our experience, we have found that the phrase "swarm" is often used interchangeably with "multi-agent systems" (which is commonly used for simulated agents) and "multi-robot system," with the exception that perhaps "swarm" is a single word that encapsulates what might otherwise be terms that are used more commonly among research communities. The word "swarm" connotes a group of several vehicles working together; the approaches that address what swarms can do have for years been the purview of academic investigations including multi-robot task allocation (MRTA). Work on swarms is also commonly attributed to the Artificial Life (ALife) community, where topics in complex adaptive systems, emergence, and self-organization are explored to understand life and life-like processes. Craig Reynolds' work on "Boids" (for "bird-oid objects") in 1987 demonstrated swarms of agents that fly in formation by following three simple rules: a separation rule that keeps agents a certain amount of space apart from each other, an alignment rule that keeps agents in the same general direction as other agents, and a cohesion rule that keeps agents moving towards a center of mass with the other agents. Combining these three rules into a single vector to indicate each agent's movement results in an interesting visual display of flocking behavior. This behavior, an emergent property of three simple rules, has stoked imaginations and is likely a reason for the prevalence of the term "swarm" in the general population today.

Based on our experience, the word "swarm" may be used for three or more vehicles (the idea that the vehicles are mobile is another assumption of the word "swarm"), where the number of vehicles depends on the application. For example, a swarm of three to five vehicles could fly ahead of trains curving around rock-strewn mountains or vehicles that are particularly difficult to maneuver to provide early warning of obstacles ahead. Around five vehicles can work together to explore and measure a planetary surface. A swarm of five to ten vehicles could be used in a military context to understand the complexities of an area of operation before additional forces arrive. A swarm of fifty to two hundred vehicles can provide surveillance of an urban area of operation, and subgroups from such a swarm could be called on to provide assistance for specific challenges or to closely inspect areas of concern.

So far, we have considered swarms that are composed of flying vehicles, but the future potential of swarms needs to consider that there are several types of vehicles that can coordinate their activity. For example, in an agricultural setting, a swarm of vehicles may include UAVs to detect areas where insect damage appears to be greatest or hydration seems to be an issue, ground vehicles that are the roboticized tractors and harvesters of the future, and even sessile sensors that are incorporated into irrigation systems. Among aerial vehicles, quadcopters provide a different kind of performance, maneuverability, and range than fixed-wing craft. Among planetary exploration vehicles, only wheeled or treaded vehicles can be used in environments that have little or no atmosphere, such as the Moon. A swarm of underwater vehicles in cooperation with aerial vehicles could coordinate their activity to count marine life or detect adversarial watercraft. Snake-like robots can explore crumbled buildings that have fallen from earthquakes or bombings. In each of these cases, the differences in vehicles' maneuverability needs to be accounted for in task allocation; each vehicle will have a different ability to traverse a region, they will provide differing amounts of speed at which they can cover the area, and they will consume their battery charge differently.

In addition to heterogeneity in maneuverability, vehicles can also differ in the ways that they interact with the environment. Vehicles may be equipped with different sets of sensors, including expensive and heavy sensors (such as a radiation detector) that may be more rare in the swarm compared to commonly-available video processing sensors that each vehicle may possess. Vehicles may have different types of effectors, as might be the case with robots on a planetary surface where some may dig the surface, some may transport the dug material, and some may process the material.

One type of heterogeneity that is starting to be explored more in research is the concept of vehicle health. As vehicles perform their work, they will start to deteriorate. The most common and benign type of deterioration is simply the amount of time that a battery-powered vehicle can operate on a single charge, and the cost of certain activities on a vehicle's battery level. Other types of deterioration include dust that can get into a vehicle's wheels or cover a vehicle's solar panels, damage incurred by operating in a dangerous environment such as a collapsed building, compromised sensing capabilities, and other kinds of wear.

How these types of heterogeneity affect the swarm and its operation is managed by the swarm's task allocation capabilities.

TASK ALLOCATION

Task allocation is the means by which the swarm determines how to accomplish the objectives set forth by the human operators. While a swarm is comprised of individuals, a swarm is an entity of its own, and ought to be treated as such. If a human operator were to control each robot instead of the swarm, then the operator isn't using a swarm, they are just using a bunch of robots. A swarm implies that there is coordination among the members so when the human operator directs the swarm to "find survivors in the rubble" or "ensure this field will yield a productive crop," the swarm will determine how to make those objectives happen. This is especially important for swarms that are out of communication range of their human operators. For example, if a swarm were to explore the surface of Mars, the amount of time it takes for a message to get to and from Earth may be too long for the swarm to make a quick decision in response to an issue in the Martian environment. How the swarm distributes tasks to the individual members is the concern of task allocation, and multi-robot task allocation (MRTA) is an active area of research that tends to converge on several algorithmic techniques.

One of the first questions to address in task allocation is to understand whether the swarm will operate in a centralized or distributed (or decentralized) manner. In a centralized swarm, a single entity (which can be a member of the swarm, or a computer outside of the swarm that directs the vehicles) determines which vehicles will perform which tasks. Task allocation algorithms often require a significant amount of computational power, and having a single system that can determine the tasking for each vehicle makes sense when the system making the determinations is computationally powerful and when communications between the centralized system and the individual vehicles can be ensured. The risks of centralized task allocation are a single point of failure, and the need for constant communication with the central system. In decentralized coordination, each vehicle has its own instance of the task allocation algorithm, and each vehicle determines its own actions based on what it knows about itself and about other vehicles in the swarm. This kind of coordination requires that each vehicle is computationally sufficient to run the algorithms, and that each knows about the other vehicles in the swarm. The benefit of decentralized task allocation is that any vehicle in the swarm can leave the swarm (whether intentionally or not) and the remaining vehicles can still operate.

Common algorithmic techniques in MRTA include integer linear programming (ILP), partially observable Markov decision process (POMDP), and linear temporal logic (LTL). MRTA algorithms are different from swarm intelligence algorithms, which use the concept of swarms as inspirations for algorithms that explore a space of possible solutions to a problem. For example, particle swarm optimization (PSO) and ant colony optimization (ACO) are two types of swarm intelligence algorithms that use agents to search a space of solutions for an optimal solution to a problem. Regardless of the MRTA algorithm that powers the task allocation, considerations must be made for the dynamic nature of real environments. Unlike laboratory environments in which algorithms can be explored by how well they allocate tasks at a single point in time under ideal conditions, in applied settings, the environment is constantly changing: new hazards or conditions come into play, including weather conditions or adversary actions. Vehicles may fail or be destroyed, or new vehicles may join the swarm. While an MRTA algorithm may determine an optimized task allocation under an initial ideal set of conditions, the changes experienced in real environments means that a continuous understanding of the environment and a re-assessment of the allocated tasks in response to those changes needs to be built into the real-world system.

Heterogeneity of the vehicles must also be accounted for in task allocation, and this includes the changing nature of the health and status of vehicles. And, recalling the discussion above about emergence from simple rules, task allocation needs to account for both intended and unintended emergence from the interaction of simple rules. Techniques such as Monte Carlo simulation can be used to evaluate the task allocation algorithms against multiple scenarios, and generative adversarial networks (GANs) can be used to attempt to thwart task allocation algorithms to determine their weaknesses so researchers and developers can account for such weaknesses before swarms are deployed to the real world.

COMMUNICATION

One of the challenges in the real world is the availability of communication between vehicles in the swarm. First, the size of the swarm has an impact on communications. For a small swarm in a well-known environment, it is feasible for each vehicle to keep in touch with each other vehicle. In larger swarms, the amount of bandwidth available may be unable to sustain the number of communications necessary between vehicles (for an n-vehicle swarm, up to n^2 communications channels may be necessary in a naïve communication implementation), and the amount of computational power necessary for each vehicle to process the incoming messages may be too great for the vehicles to handle. In real environments, communications could be spotty because of geophysical limitations or active denial by an adversary. If communications cannot be guaranteed, how can vehicles ensure that they operate without conflicting with other vehicles? How can human operators be assured that all vehicles in the swarm will respond to human control?

Techniques for managing communication limitations can be incorporated into task allocation (especially decentralized task allocation). Communication among swarm vehicles in proximity may be prioritized over communication with distant vehicles; for example, it is more important to know what your immediate neighbors are doing so you can coordinate closely, and it is less important to know what vehicles a mile away are doing, unless those vehicles are engaged in some kind of coordination that needs to tie their activity together, such as detecting and then responding to illegal cargo being transported on a highway.

Techniques to mitigate communication limitations can maximize the performance of the swarm despite communication difficulties. In a communications-denied environment, vehicles that are close to each other may still be able to communicate, so a neighbor-centric approach where each vehicle can communicate with those that are physically close may be used. Passing along communications can be used to bridge distance, and vehicles that serve as repeater nodes can be deployed in areas where communications in a physically restrictive environment, such as a building collapse or underground area, is difficult.

Insect societies that serve as an inspiration for swarm work employ another type of communications modality called stigmergy, which is the use of markers in the environment as a communications mechanism. For example, ants and termites place pheromones in the environment, and these are cues for other ants and termites as they work together to forage for food or build a nest. Swarms exploring a collapsed building could similarly use something like pheromones – perhaps colored spray paint – as an analogue to pheromones to mark which pathways have already been explored. While stigmergy is commonly found in nature among eusocial insect societies, it is relatively unexplored in physical swarm systems.

SWARM COORDINATION FRAMEWORK

At Charles River Analytics, we have explored these challenges through our work on applied swarm systems for customers including DARPA and NASA. As a function of this work, we have developed the Swarm Coordination Framework (SCF), a software capability that lets us understand and evaluate the performance of simulated and real-world swarm systems. Our focus is on heterogeneous swarms that use either centralized or decentralized task allocation to determine vehicles' assignments. As a framework in the truest sense of the word, with SCF we are able to experiment with and replace specific subsystems of the swarm, such as the task allocation algorithms or the communication approaches, as well as the environment itself, while keeping the other aspects of the swarm system constant, which provides a compelling capability to evaluate and assess the effectiveness of multiple approaches to swarm systems.

We have developed SCF to deploy to physical vehicles, which means that the behaviors we assess in simulation are the same behaviors that we assess when the software is placed on robots. This is possible through an abstraction layer that we have built into SCF, which realizes that swarm behaviors are purely logical concerns that depend on, but are not specifically tied to, physical aspects of the vehicles, such as controlling the vehicle's movement, reading its sensors and activating its effectors, and using its communication system.

We are performing unique research in advancing the notion that vehicles are heterogeneous not just in their mobility and sensing/effecting but also in their health and status. With this comes our unique extensions to task allocation algorithms that incorporate vehicles' health and status, and the ability to probabilistically ascertain the future health and status of vehicles and to use that information in allocating tasks. For example, a vehicle can anticipate the future battery level of another vehicle if it were to go to a distant location to performing a sensing activity and then return, and then consider how the vehicle's battery status upon its return would affect its ability to perform other tasks. That information may be used in conjunction with other facts about the swarm to determine which vehicle should actually be assigned the sensing task.

We believe in the significant innovation that swarm systems can provide in areas such as smart agriculture, search and rescue, environmental monitoring, space exploration, and military operations, and we welcome opportunities to work with industry and military leaders to continue to make dependable and trustworthy swarm autonomy a reality.

MAKING THE BEST OUT OF YOUR FLIGHT BATTERY: HOW TO WOW CUSTOMERS AND REGULATORS

Ryan Spray, Ph.D.,*

Electrification of flight and Urban Air Mobility can reduce carbon emissions and pollutant emissions that disproportionately effect lower income citizens who live near significant traffic and congestion. Electrified flight will be assessed differently (than traditional engines) by the FAA in certification. Battery technology may limit progress and/or viability of electrified flight as a business. The landscape of battery power for UAS / flight, as well as ground applications, has become crowded and confusing at times. This paper will highlight both the opportunities and challenges of using lithium-based battery technologies for flight, as well as testing/standards and regulatory aspects of their use.

INTRODUCTION

The proposed benefits of electrified air mobility are many, and frequently tied to the demand for a greener economy. Supporters suggest benefits ranging from improved last-mile delivery services to less congested freeways and roads, and agencies such as National Aeronautics and Space Administration (NASA) and the Federal Aviation Administration (FAA) have added credibility by partnering to "accelerate the realization of emerging aviation markets for passenger and cargo transportation in urban, suburban, rural, and regional environments" through targeted campaigns[†]. North America and Europe are positioned as forerunners of a multibillion-dollar global market for Urban Air Mobility (UAM) alone[‡].

Due to the need of high energy density power sources, aircraft have traditionally relied on hydrocarbon fuels. Advances in motor and battery storage technology over the last 20-30 years have enabled battery-powered vehicles to become preferable in applications where attributes such as low noise, no liquid fuel management or exhaust fumes, minimal vibration, short torque rise time, or minimal maintenance are important aspects of said application. While liquid hydrocarbon fuels currently have higher energy density than batteries when compared in a vacuum, the combined gravimetric and volumetric energy density of a battery / motor system can often exceed the requirements for a given device or application. Fuel cells offer many of the benefits of battery-electric powered systems. Additionally, fuel cells / electrolyzer systems can provide for remote generation of hydrogen from water sources. However, traditional fuel cell technologies do generally suffer from durability and reliability challenges in comparison to battery powered systems, particularly in mobile applications.

In general, the extent that electrification / batteries will play a role in the overall flight operations will depend on the power / energy demands of the application, which does not currently make them

* Principal Scientist, Polymer Science & Materials Chemistry, Exponent Inc., 1075 Worcester St. Natick, MA 01770
† https://www.nasa.gov/aamnationalcampaign
‡ https://www.prnewswire.com/news-releases/5-23-bn-urban-air-mobility-uam-market-2035--300863223.html

good candidates without further technology advancements. This paper will describe the state of the art for battery technology, as well as both technical and regulatory hurdles that remain.

Figure 1. Electric aviation energy storage R&D needs, from US Department of Energy* (2019).

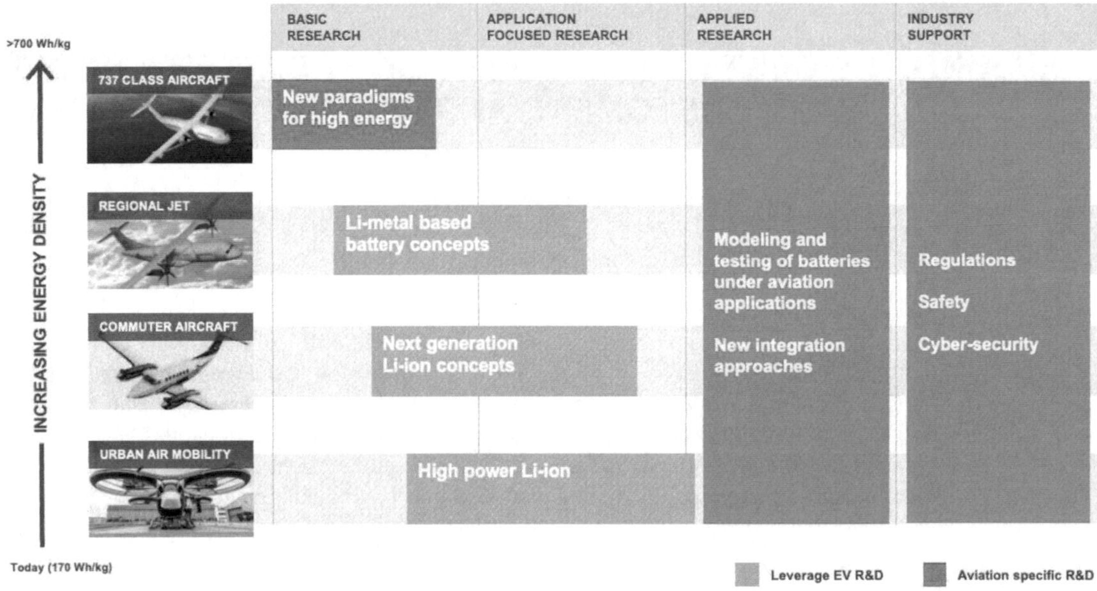

BACKGROUND ON LITHIUM-ION BATTERY TECHNOLOGY

The term lithium-ion (Li-ion) battery refers to an entire family of battery chemistries. It is beyond the scope of this report to describe all of the chemistries used in commercial lithium-ion batteries. In addition, it should be noted that lithium-ion battery chemistry is an active area of research and new materials are constantly being developed and new cells are being engineered. This chapter provides an overview of the technology and focuses on the characteristics of lithium-ion batteries common to the majority of available batteries. In the most basic sense, the term lithium-ion battery refers to a battery where the negative electrode (anode) and positive electrode (cathode) materials serve as a host for the lithium ion (Li+). Lithium ions move from the negative to the positive electrode during discharge and are intercalated into (inserted into voids in the crystallographic structure of) the positive electrode. The ions reverse direction during charging, de-intercalating from the positive electrode and intercalating into the negative electrode. Since lithium ions are intercalated into host materials during charge or discharge, there is no free lithium metal within a lithium-ion cell.

In a lithium-ion cell, alternating layers of negative and positive electrode are separated by a porous film (separator). An electrolyte composed of an organic solvent and dissolved lithium salt provides the media for lithium-ion transport. A cell can be constructed by stacking alternating layers of electrodes (typical for high-rate capability pouch cells), or by winding long strips of electrodes into a "jelly roll" configuration typical for cylindrical cells. Electrode stacks or rolls can be inserted into hard cases that are sealed with gaskets (most commercial cylindrical cells), laser-welded hard cases, or enclosed in foil pouches with heat-sealed seams (commonly referred to as lithium-ion polymer cells, although this terminology is not accurate). A variety of safety

---

*https://jcati.org/sites/default/files/Assessment%20of%20R%26D%20Needs%20for%20Electric%20Aviation_FINAL.pdf

mechanisms might also be included in a cell mechanical design, such as charge interrupt devices and positive temperature coefficient switches, based on the intended application.

An individual lithium-ion cell will have a safe voltage range over which it can be cycled that will be determined by the specific cell chemistry. For most commercial lithium-ion cells, that voltage range is approximately 3.0 V (discharged, or 0 % state-of-charge, SOC) to 4.2 – 4.4 V (fully charged, or 100% SOC). For most cells, discharge below the manufacturer's specification (i.e. 3.0 V) can cause degradation of electrodes and is referred to as over-discharge. Repeated over-discharge, combined subsequent charging, can lead to cell failure and cell thermal runaway (discussed below). Thermal runaway is the state where self-heating produced by the sample, a lithium-ion cell in this case, occurs more quickly than heat can be shed to surroundings, leading to uncontrollable temperature rise and catastrophic failure of the cell. For most cells, significant overcharge can lead to rapid, exothermic degradation of the electrodes. Since overcharging can lead to violent thermal runaway reactions, a number of overcharge protection devices are either designed into cells or included in the electronics protection packages for lithium-ion battery packs.

A lithium-ion battery (or battery pack) is made from one or more individual cells packaged together with their associated protection electronics. By connecting cells in parallel, designers increase pack capacity. By connecting cells in series, designers increase pack voltage. Thus, most battery packs will be labeled with a nominal voltage that can be used to infer the number of series elements and pack capacity in Ampere hours (Ah) or Watt hours (Wh) that will provide an indication of the capacity of each series element (size of individual cells or the number of cells connected in parallel). For large format battery packs, cells may be connected together (in series and/or in parallel) into modules. Large format battery pack architecture can be significantly more complex than small consumer electronics battery packs which typically contain series connected elements consisting of two or more parallel connected cells. Nonetheless, the simplified analysis method used above can still be applied to generally understand the total number of series elements within a battery pack, and the capacity of the parallel elements.

Assembling battery packs in series and parallel to meet power requirements is not the only requirement, as battery packs for critical applications should be designed for resiliency, which includes accounting for both abuse tolerance as well as tolerance to random faults and failures. The rise of electric vehicles of all shapes and sizes has led to the design and development of more mechanically (and electrically) robust battery designs to deliver sufficient performance, reliability and safety. Regarding abuse tolerance, the design and validation must be guided by both the expected application environment as well as other foreseeable misuse(s) or environments. Typical strategies for abuse tolerance include, but are not limited to, use of DFMEA in the design process, drop/shock/vibration testing, atmospheric ingress (e.g. humidity, salt spray) testing, impact or puncture testing, and programmed electrical safeties limiting operation due to current, voltage and/or temperature limits. The breadth and details of these analyses are subject to the specific applications and often standards and regulations in specific industries.

TECHNICAL CHALLENGES FOR ELECTRIFIED FLIGHT

Energy density

To meet the allotted design space for the power system, batteries must be designed to be energy dense. This is particularly applicable for flight as their density will greatly affect the range and operational mission lifetime before the need for recharging. Energy density can be assessed in terms of gravimetric or volumetric energy density, with the merits of each varying on the intended design / application. For example, gravimetric energy density will typically be more important for flight and thrust applications, while volumetric energy density can be more important for space-constrained or micro-sized applications.

The advancement of materials processing and manufacturing technologies over the last 20-30 years has led to great increases in energy density, even while utilizing the same materials / classes. Lithium cobalt dioxide (LCO) as the positive active material remains one of the most dominant chemistries for a lithium-ion cell in consumer electronics due to its energy density. Lithium nickel manganese cobalt oxide (NMC) and lithium nickel cobalt aluminum oxide (NCA) have become the dominant positive active material chemistries in the mobility space due to their high rate capability (compared to LCO).

Graphite or hard carbon have been the standard for the negative active material in lithium-ion batteries since the early 1990's for various reasons that this document will not fully detail. Metallic lithium as the negative electrode was attempted both in R&D and for commercial applications prior to this, however safety issues arose at the time with short-circuit failures due to the inherent properties of metallic lithium re-plating during charge to form dendrites that can bridge or puncture the separator.

Due to advances in the last 30+ years of chemistry, engineering, and materials science, as well as the technology and tools required to study these systems, new materials and methods have been proposed to enable the use of lithium metal in rechargeable batteries. This has led to many start-up companies that have been formed with investment and backing from interested industry partners, primarily the automotive sector. Although the commercialization viability and timeline of these technologies is varied (and it is not the intention of this paper to ascertain either), it is likely that some of the technology developed during this 'solid state battery renaissance' will find some uses in at least niche applications seeking high energy density with lower (e.g. <500) cycle life expectations.

Fast Charging

While increasing energy density of the integrated battery packs can prolong their operational use in the field, the ability to quickly recharge the battery systems to re-deploy is an important feature. This challenge had been critical in the automotive electric vehicle (EV) industry, where fast charging enables the vehicles to be more competitive with the refueling of traditional combustion engines, and is at least equally important for aviation. Charging times will mainly depend on (a) the energy of the battery (kWh) (b) the power capability of the charger and often (c) the battery and/or environmental conditions.

Regarding the battery and environmental conditions, fast charging is more stressful to a battery than lower-rate charging. In order to reliably and safely fast charge a lithium-ion battery, the charging must be performed within the appropriate temperature / voltage / current operational window to prevent degradation and damage to the battery materials.

Many newer EV batteries (e.g. 50 kWh+) are capable of Direct Current Fast Charging at rates up to 200 - 300 kW or more, which can charge 80% or more of the capacity in <30 minutes, although it is worth notice that this comes with the requirement of significant cooling and capable infrastructure.

In general, heat generated by passing high currents within the battery can damage the active materials and create degradation products that may be irreversible. Additionally, attempting to insert lithium ions into the negative electrode during charge faster than it is physically capable of accepting the ions will lead to plating of metallic lithium (i.e. no longer intercalated ions) onto the surface of the electrode. Lithium plating can create many problems for performance, reliability and safety including electrolyte degradation, gas generation, lithium plating and associated performance loss and potential internal short-circuiting.

WHAT STANDARDS CURRENTLY EXIST FOR FLIGHT BATTERIES?

The most often-cited standard for rechargeable batteries in flight applications is the RTCA DO-311A, the "Minimum Operational Performance Standards for Rechargeable Lithium Batteries and Battery Systems." This standard provides design, testing, and installation guidance for rechargeable lithium battery systems that are permanently installed on aircraft including standalone and embedded batteries. It addresses all sizes of rechargeable lithium battery systems regardless of energy content, but allows for exemption of batteries <2 Wh provided they meet the requirements of the applicable UL or IEC standards.

Regarding performance topics covered in DO-311A, the standard covers topics such as battery capacity, elevated and low temperature capacity, capacity retention, and drop/impact resistance, as well as management / protections such as short-circuit protection, over-discharge protection. Safety-related performance requirements also exist, including containment for single-cell thermal runaway, battery-level thermal runaway, as well as explosions (containment). It is important to ensure understanding and compliance of all the requirements. It is also valuable to understand battery performance and standards for comparable or adjacent industries to understand performance benefits and limitations of available batteries for flight applications.

Standards should be treated as a best first step toward ensuring a battery system is right for the desired application. Application-specific testing is also necessary to understand how the battery system operates outside of very specialized conditions that may have little overlap with real-world operation. Performing a Design Failure Mode and Effect Analysis (DFMEA) is a good way to understand the expected performance as well as gaps that exist between known performance and nominal conditions and anticipated or foreseeable abnormal conditions.

HOW WILL THE FEDERAL AVIATION ASSOCIATION (FAA) REGULATE UAM AND OTHER ELECTRIFIED FLIGHTS?

Unmanned aerial vehicles (UAVs) provide a good example of how regulations are likely to play out for Urban Air Mobility (UAM). In 2016, the FAA made it possible to operate commercial UAVs that comply with current standards and current pilot training requirements[*†‡]. Since then, additional regulations have been added that clarify or expand initial policies, such as remote identification rules in 2019, and the ability to fly drones at night in 2021. This appears to bode well for manned electric vertical take-off and landing (eVTOL) aircraft. The FAA has indicated that it is actively preparing for a future that integrates electrified UAM into aviation infrastructure—but it's important to keep in mind that new aircraft designs will require new certification standards both for pilots and the aircraft itself. As another example, manned vehicles have very different risk profiles from unmanned vehicles, thus very different regulations, which makes sense for UAM, too.

Outside of FAA regulations, another key point is that while National Airspace System (NAS) is regulated by the FAA, takeoff and landing can be regulated locally. This may create a patchwork system where heliports (or vertiports) are prevalent in some cities while absent in others. For instance, today there are only three heliports in New York City compared to dozens in Los Angeles. Whereas New York has a track record of caution, helipads were required in Los Angeles on buildings over 75 feet for decades because of their value to emergency services and fire response. Citizen complaints eventually led to the law being rescinded in 2014. These same dynamics are certain to

---

[*] https://www.reuters.com/article/us-usa-drones-faa/u-s-to-allow-small-drones-to-fly-over-people-and-at-night-idUSKBN2921R8

[†] https://www.faa.gov/newsroom/faa-announces-effective-dates-final-drone-rules

[‡] https://www.nytimes.com/2016/06/22/technology/drone-rules-commercial-use-faa.html

impact the future of electrified UAM, as the tolerability of the public toward environmental impacts, noise, and safety will be a powerful factor in how local authorities proceed.

CONCLUSION

While electrified flight in some form appears to almost inevitable in the future, the size and types of aircraft that become a reality still seems heavily dependent on technical challenges and regulatory hurdles. Frameworks for flight certification of batteries exist, however there remains uncertainty on how certification for battery-electric flight propulsion should differ, particularly for passenger flights. The testing and analyses involved in battery flight certification are an important first step in determining airworthiness, however application-specific testing and analysis is necessary to understand how the battery system could operate in real-world operation. Technologies exist to make some levels and ranges of electrified flight possible, however advances in battery technologies are required to reach energy densities that enable larger payloads and range and it is important that these technologies are vetted thoroughly for the application.

ACKNOWLEDGMENTS

Thanks to Adam Dershowitz, Ph.D., P.E., CFEI for discussions and input regarding aviation and Urban Air Mobility.

LIMITATIONS

This paper contains material to be presented during oral presentations delivered on April 27th, 2022 at the AUVSI Xponential Conference. The material contained herein is presented to a reasonable degree of scientific and engineering certainty, and may not adequately address the needs of users of this presentation. Any re-use of this presentation, or any of its contents, is made at the sole risk of the user. No guarantee or warranty as to future relevance is expressed or implied. Exponent reserves the right to supplement this presentation and to expand or modify its contents based on review of additional material as it becomes available and/or through any additional work or review of additional work performed by others.

USING RADAR FOR AUTOMATED DETECT AND AVOID MANEUVERS AGAINST NON-COOPERATIVE UAVS

Lawrence J. Surace*

The air-to-air detection, track establishment and execution of a full autonomous collision avoidance maneuver, especially against a small non-cooperative fast-moving target, low altitude target, is a challenging one indeed. Yet this is exactly the challenge Honeywell Aerospace has taken on to establish the operation of its airborne detect and avoid system. At the core of this capability is a small, light-weight yet very capable airborne radar, the RDR-84K. The nomenclature, RDR-84K, embodies a radar of approximately 8" by 4" operating in the K frequency band. The RDR-84K is a full phased array radar capable of receiving multiple beams within a large air space volume. This paper outlines the architecture, implementation and results of a comprehensive flight test campaign conducted by Honeywell Aerospace in 2021 to demonstrate this capability.

INTRODUCTION

In 2021 Honeywell Aerospace embarked on a three-step flight test campaign aimed at developing, refining, and demonstrating the Detect and Avoid capability of the RDR-84K radar system, those steps are summarized below.

Step 1: Integrate the low Cost, Size, Weight and Power (C-SWAP) RDR-84K radar onto a small UAV platform practical enough to support a high temp flight test campaign.

Step 2: Characterize the radar's low altitude (<400' AGL), high clutter performance incorporating the use of a manned (Robinson R44) helicopter as the initial intruder vehicle.

Step 3: Fully demonstrate the low altitude autonomous Detect and Avoid guidance maneuver against a low radar cross section, non-cooperative intruder vehicle, a DJU M600 drone.

Version 1.0 of the RDR-84K is the development model of the radar and has undergone significant testing on several platforms, to date most of that testing has been at higher altitudes (above 400 AGL). The purpose of this flight test campaign is to test the radar's clutter rejection capability at low altitude. The RDR-84K radar is a K-Band air to air, air to ground, ground to air, phased array radar capable of both azimuth and elevation electronic beam steering and stabilization. Figure 1 below provides the full specification of the radar.

* UAS / AAM Architect, Honeywell Aerospace, 21111 N. 19th Ave, Phoenix Az. 85027

RDR-84K V2.0 SUMMARY (IN DEVELOPMENT)

Parameter	Specification
Modulation Type	Frequency Modulated Continuous Wave (FMCW)
RF Operational Frequency Range	24.45 - 24.65GHz
Size, Weight	8.9" x 4.9" x 1.4", < 1.5lbs
Power Required	11-36Vdc @ 65Watts
Transmit power	Approximately 0.5W into 17dBi antenna aperture
Field of Regard	Azimuth = 110° / Elevation = 30°
Antenna Beamwidth	RX = 8° x 8° TX = 8° x 110°
Angular Resolution	<1°
Range Resolution	Better than 1 meter
Cooling	No cooling required
Operating Temperature range	-55° to +70° C ambient

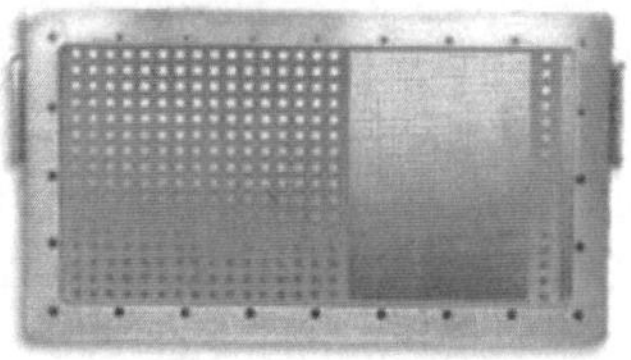

Figure 1. RDR-84K Specifications

Integration of the radar onto a viable UAV test platform required taking into consideration several factors such as.

1. Platform robustness: the UAV platform had to be capable of lifting a significant payload (>10 lbs., exclusive of batteries and flight control system) and have a wide flight envelope to include tolerance of cross wind conditions more than 15 knots.

2. Powerful, interchangeable, and long-lasting batteries (capable of supporting flight times exceeding 20 min on a single charge) were required.

3. Realistic flight speeds: Honeywell requirement was for the UAV to be capable of speeds greater than 15 m/sec.

4. Integration with a highly robust and programmable mission planning software system to allow automatic, consistent, and repeatable flight mission profiles and post flight analysis.

5. Highly reliable and robust ground communications and control platform allowing a Part 107 pilot "in and over the loop" flying, was defined as a critical safety capability when flight testing autonomous functionality.

In compliance with these requirements Honeywell selected the SE-8 quadcopter from BFD systems. This was done to provide a more realistic flight profile model of a radar platform and represented the most expedient approach for Honeywell to meet the program schedule.

STEP 1: INTEGRATION OF THE RADAR ONTO A TEST PLATFORM

Since Version 1.0 of the RDR-84K was used in this flight test, an NVIDIA Jetson Xavier storage unit with 1TB of data storage was installed to provide additional data storage and serve as the host DAA processing platform. The final Honeywell DAA system incorporating the production Version 2.0 of the RDR-84K will have additional storage and processing capability built into it and therefore a co-processor will not be required.

The BFD drone was supplied from the factory with the Orange Cube flight control system installed. The IMU output and GPS position / timing was used for "ownship" position information and timing provided to the radar to allow the beam stabilization software to compensate for vehicle motion.

Real-time data communication to a ground control station was established through two 2.4 GHz Microhard Digital Data Link devices. These devices proved to be highly reliable and rugged while consistently providing real-time communications to the drone exceeding 2 km ranges. The ability of this real-time link is extremely valuable ensuring both safety of the vehicle and monitoring real-time radar returns at the base station.

A uAvionix ADS-B receiver was also integrated into the architecture to allow detection and displaying of cooperative vehicle within the airspace and to provide situational awareness to the ground station. This allows the BFD drone pilot to observe cooperative vehicles in the airspace and a secondary observer to monitor the local airspace for any potential unwanted aircraft intrusions during the controlled flight testing. Figure 2 below outlines the detailed architecture integrated onto the BFD platform.

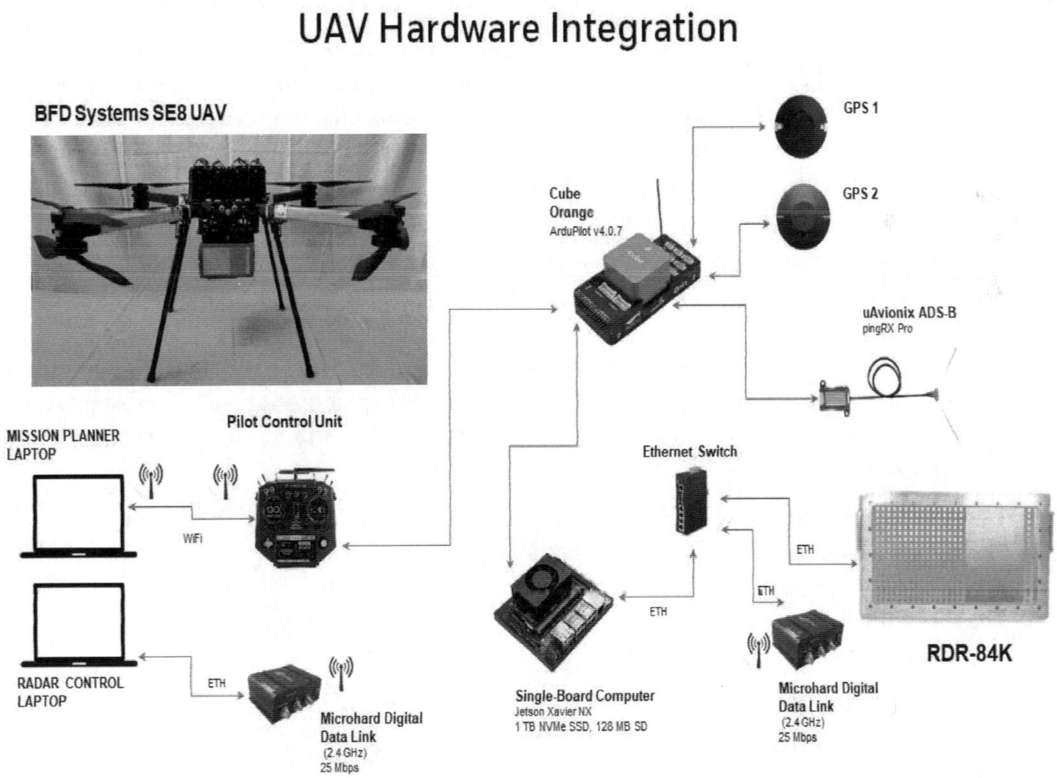

Figure 2. UAV DAA Hardware Architecture

Figure 3 below shows the final implementation of the radar system payload on the BFD drone

Figure 3. DAA Architecture Fitted onto Test Platform drone

STEP 2: CHARACTERIZATION OF THE RADAR'S LOW ALTITUDE PERFORMANCE

To perform this testing of the radar's low altitude / high clutter performance, Honeywell contracted with a local helicopter operations firm in Chandler, Arizona to supply the R44. A Honeywell helicopter pilot was trained to fly the specific maneuvers. Due to the low altitude (below 400' AGL) and, in some cases high speed flight, specific engine out and low altitude autorotation training was conducted to ensure a high degree of safety of the flight test campaign. A total of 20 hours of this flight training was conducted over a one-month period.

A critical program consideration was the selection of an appropriate location to conduct this low altitude manned flight test campaign. Factors such as safety, airspace density and control were key elements considered. By its location, population density and ideal flying conditions for much of the year, the Phoenix Metroplex airspace is very crowded and therefore difficult to conduct flight testing of this nature. Given these parameters Honeywell, in collaboration with Aerospace Arizona (a state sponsored economic development organization) was able to secure a contract with Benson Municipal Airport to allow this flight testing.

Benson is located 50 miles southeast of Tucson Arizona, is an uncontrolled airport with a 4000+' marked and paved runway, ideal for the alignment parameters required for these tests. Additionally, under this arrangement, Honeywell (working with the Benson Airport Operations) was able to secure flight times and airspace access controlled by NOTAMs to ensure no other aircraft were operating in the area during flight testing.

Honeywell focused on refining additional program specifics outlined below:

- Primary objective to evaluate the operation of a single radar panel.
- Establish this as a baseline for performance.

- Single panel operation will be evaluated for 120 and 110-degree azimuth field of view performance.
- All operations will be conducted at 300 – 400 feet AGL.
- The Robinson R44 as the intruder target.

The configuration of the various helicopter (intruder) profiles overlaid on the airport is shown in Figure 4 below. The required distance for the BFD radar drone starting from point O to point M is outlined along with each of helicopter flight profiles. Prior to the actual testing Honeywell conducted several test flights at the Benson airport to ensure proper alignment and acceleration times required for the drone. During these times an appropriate distance was added to Point O location to accommodate acceleration times of the drone to the required 20 m/sec.

By doing this it was assured the radar drone was at full speed (20 m/s) at Point O and remained at that velocity until the 100 meter no transgression zone at prior to Point M. It should also be noted radar data recordings were started the instant the radar drone started its inbound run approximately 15 seconds prior to reaching Point O to capture additional data and to ensure the full distance run from Point O to the 100 meter non transgression zone was recorded.

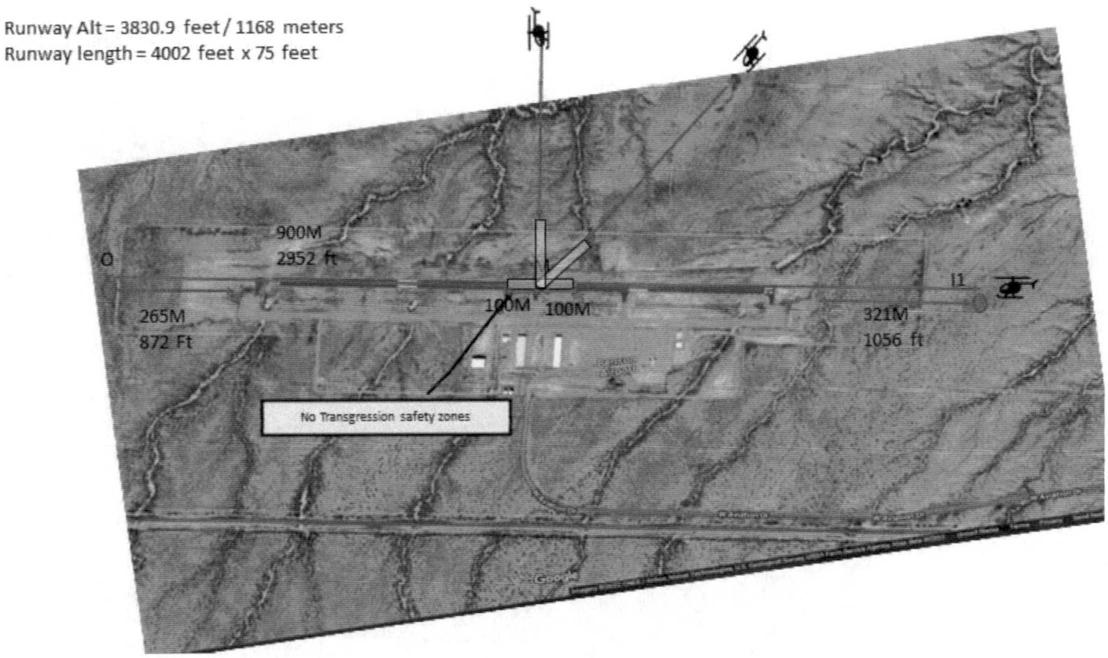

Figure 4: Helicopter / UAV Flight Paths

Helicopter Flight test parameters

This section describes the flight test parameters and test cards used by the R44 intruder helicopter to accurately conduct the flight tests. Three primary test cards were developed, one for each of the intercept angles, 180 degrees, 135 degrees and 90 degrees. These were given specific identifier names to simplify communications, the 180-degree intercept angle was called out as test path "Romeo", since it took the flight path directly along runway, the 135-degree intercept angle was referred to as test path "Victor" since its flight path took the shape of a V, and the 90-degree intercept angle was referred to as test path "Lima" since its flight path took the shape of an L.

Based a-priori information (obtained from rehearsal fights) the helicopter crew would call out (via VHF radio communications) time to the 100 m "no transgression zone" (in this case defined on the test cards as "RGAP", "VGAP" and "LGAP") for each of the test speeds. These time call outs would then trigger the drone pilot to start the radar drone inbound run to Point M, allow the drone to come up to the required 20 m/s and coordinate with the intruder helicopter to the radar convergence area. At the same time the ground station radar controller would start recording radar data until the convergence points where the test was concluded, and helicopter proceeded to its prescribed outbound leg

To enhance safety, specific "No Transgression Zones" were established 100 meters around Point M. The boundaries of these No Transgression Zones were referred to as "RGAP", "VGAP" and "LGAP" indicating from the helicopter perspective the "Go-Around Points" for each of the Romeo, Victor and Lima flight profiles. Figure 5 below shows further details of the predetermined "no transgression zones".

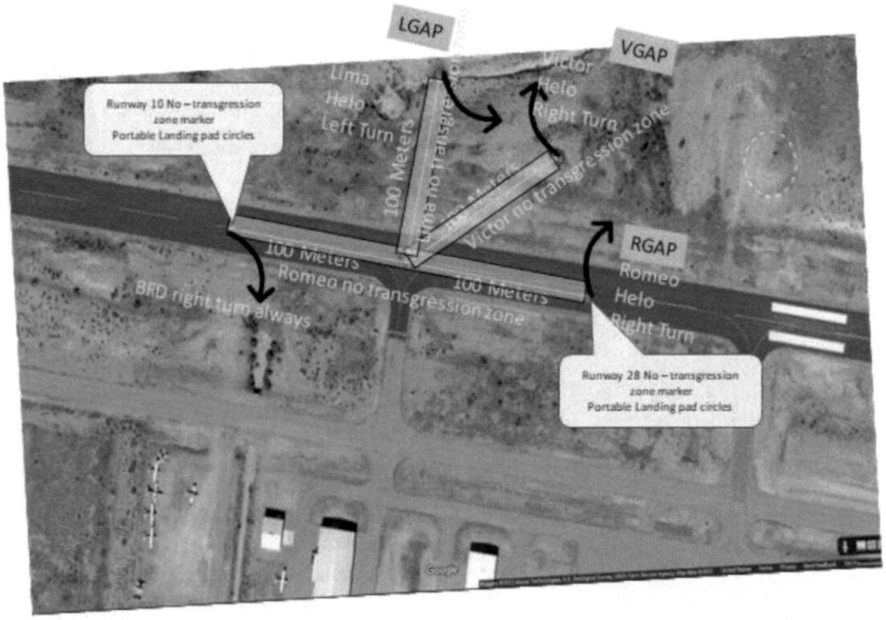

Figure 5. Helicopter No Transgression zones

Radar Performance

For non-cooperative sensors that don't internally correlate measurement returns and provide sensor specific IDs, Honeywell developed the Honeywell Correlator System (HCS) to generate sensor measurement tracks. This sequence of measurement returns and corresponding sequence ID constitutes a measurement track for radar users. HCS algorithms reside on the RDR-84K processor and not the DAA processor so that the HCS functions don't affect DAA functionality.

The HCS is based on a unique application of a multiple hypothesis testing approach like the algorithm used by Honeywell in the Honeywell DAA tracking system. The objectives of the HCS are as follows:

1. Associate and statistically fuse multiple returns originating from the same intruder aircraft at the same time epoch so that there is one measurement return per intruder aircraft at each time epoch.

2. Identify a sequence of measurements originating from the same intruder aircraft over multiple time epochs.

3. Coast through RDR-84K antenna scan pattern strategies.

4. Coast through missed detections.

5. Assign an ID to the sequence of sensor measurements.

6. Identify an unknown number of intruder aircraft.

7. Reject clutter measurements; and

8. Use range rate measurements to assist with data association when available.

It is important to note the objectives of the HCS does not include smoothing of the sensor measurement tracks, however, some smoothing of the measurement returns is inevitable for the following reasons. First, filtering is required to merge multiple returns of the same intruder aircraft at a particular time epoch. Second, filtering is required to coast through RDR-84K's antenna scanning patterns. Third, filtering is required to coast through missed detections. Fourth, filtering is required to predict measurement locations of current tracks to associate with current measurement returns.

The HCS can be divided into three major functions: Format Measurements, Correlator, and Format Track Output Statistics (Figure 6).

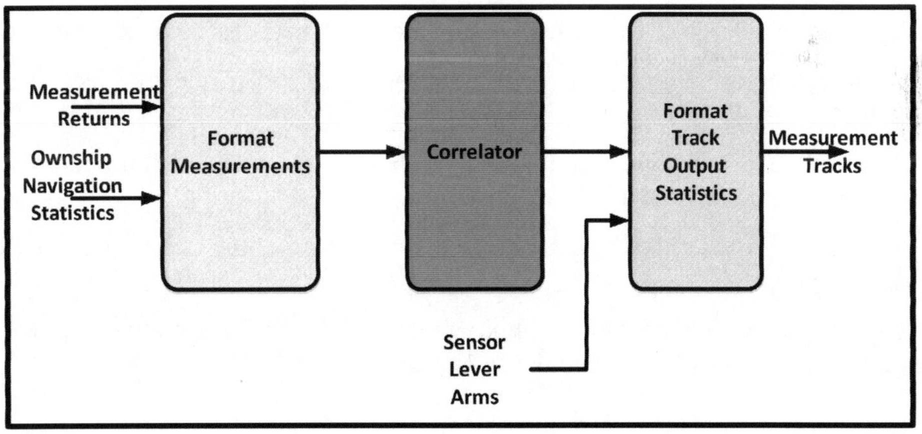

Figure 6 Honeywell Correlator System

STEP 3: LOW ALTITUDE NON-COOPERATIVE, AUTONOMOUS DETECT AND AVOID FLIGHT MANEUVER TESTING

This phase of the program focused on applying the results of the first two steps (flight test platform integration and radar characterization) toward demonstrating a fully autonomous detect and avoid maneuver against a low altitude non-cooperative intruder.

A key step was to select an appropriate non-cooperative intruder vehicle. To that end a DJI M600 (shown in Figure 7) was selected since it was determined to be representative of a realistic low radar cross intruder. The M600 also came with an all-inclusive mission planner package allowing the flight test team to generate deterministic and repeatable flight plans. Additionally, it has a comprehensive GPS and data recording capability ensuring replication of true vehicle position throughout the entire flight profile.

INTRUDER DRONE OVERVIEW

- **Incorporated a COTS DJI M600 hex copter.**
  - Small RCS
  - Non-cooperative
  - Highly maneuverable
  - Comparable flight time of BFD radar drone (25 min)
  - Able to program repeatable flight paths via available mission planner.
  - Reliable GPS based positioning

Figure 7, M600 Intruder Drone

Detect and Avoid Flight Test Objectives:

The flight test plan designed for the fully autonomous DAA was focused on the following goals.

- Conducted at low altitude (300' AGL) to fully test the radar's clutter rejection characteristics.
- Fly co-altitude, head on intruder and radar drone flight paths, both drones were flying at 5m/sec, with a closure rate of 10 m/s.
- Test the DAA algorithm's ability to autonomously steer the radar drone to a calculated avoidance heading.
- Test the ground control station's ability to display the non-cooperative intruder drone and the calculated Resolution Advisory avoidance maneuver.

Detect and Avoid Flight Test Plan:

Figure 8 below outlines the flight test plan used to demonstrate the non-cooperative autonomous DAA testing. In this test, as the starting point, the radar drone was flown to an altitude of 300' AGL directly above the ground station and pointed to a westerly heading of 270°. At the same time the intruder drone was flown to a starting point approximately 1400' away from the ground station at 300' AGL and positioned to point to an easterly heading of 090°.

It should be noted due to the compressed range size and the desire to capture these tests from the ground the ACAS-sXu algorithm used was modified to allow closer than normal flight paths and avoidance distances. For full certification purposes the ACAS-sXu algorithm will incorporate avoidance distances in compliance with the approved standards.

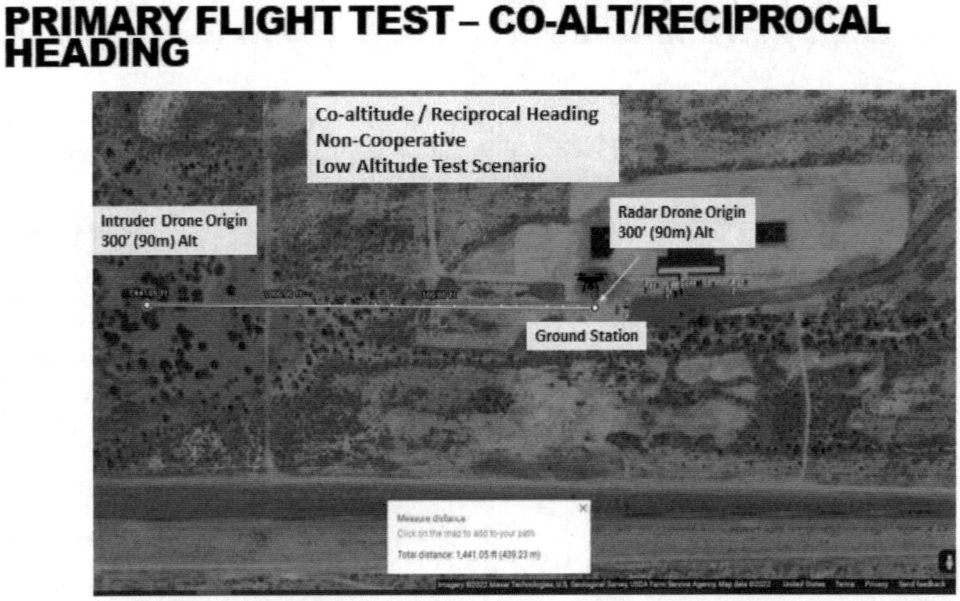

Figure 8, Detect and Avoid Radar and Intruder Drone flight paths

Detect and Avoid Auto Guidance Maneuver Results

In Figure 9 below the upper left-hand chart represents an exocentric view of the two vehicles at the very beginning of the flight test. This data is compiled from the actual flight / ground truth data retrieved from both the intruder and radar drones log files (GPS, altitude, speed, headings). The lower right-hand image is that of the ground station DAA display. As can be seen, the intruder drone symbol (taken from tracks generated by the RDR-84K) is displayed, at this point no resolution advisory has been issued or active.

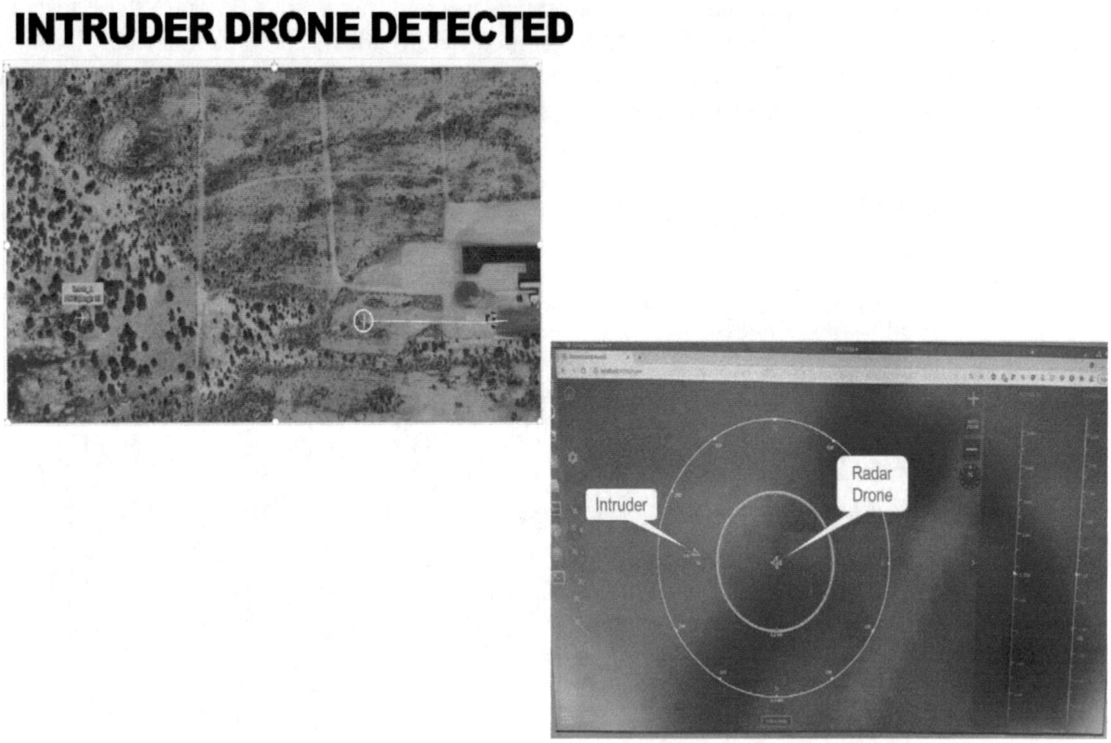

Figure 9, Start of Auto Guidance Flight – Intruder Drone Detected No Resolution Advisory

In Figure10 below the upper left-hand chart shows the truth data position of the two drones nearly mid-way through the flight test, in this case the radar is now fully detecting the intruder drone and generating tracks that are consistent with the intruder's position and sent to the DAA (ACAS-sXu algorithm). The lower-left hand chart is the DAA display output showing the ACAS-sXu algorithm calculating an avoidance maneuver heading. This calculated maneuver heading is then sent directly to the radar drone's (BFD SE8) flight control computer where its pre-programmed flight plan is interrupted with this avoidance heading.

RESOLUTION ADVISORY ISSUE AND AVOIDANCE MANEUVER STARTED

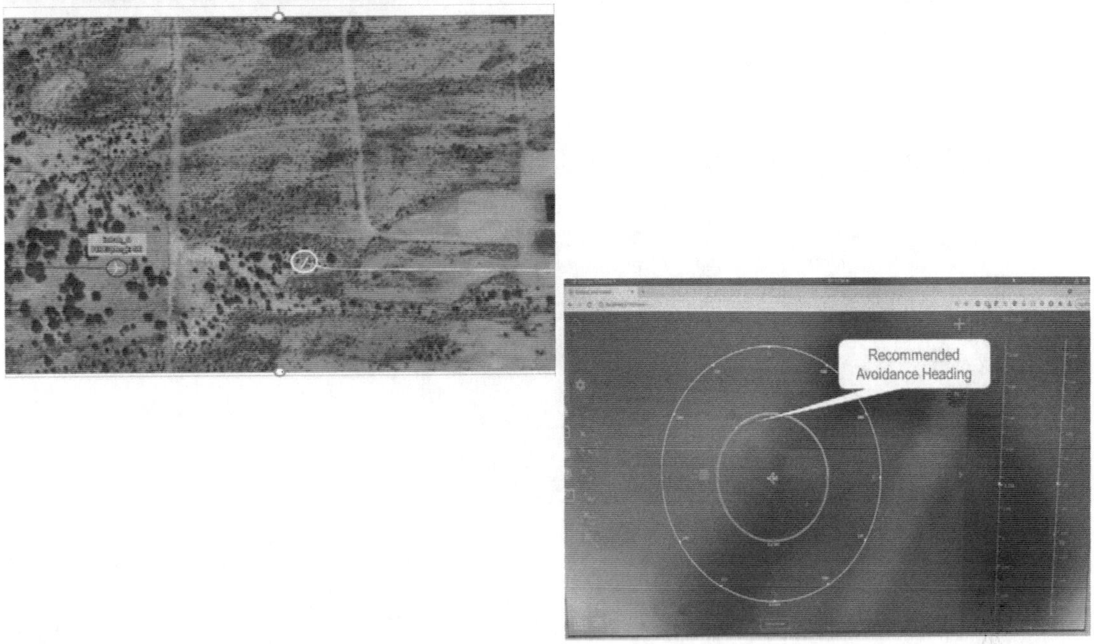

Figure 10, Midway through Auto Guidance Flight – Intruder Drone Detected Resolution Advisory Issued

AVOIDANCE MANEUVER

Figure 11, Autonomous Avoidance Maneuver executed

Since the radar drone is fitted with a single radar panel, when the radar drone executes the avoidance maneuver, the intruder drone will move out of the radar's field of regard and thus the radar will stop generating detections and tracks. Therefore, a "coast" time of 10 seconds has been injected into the detection algorithm, in this time, the tracks from the radar are discontinued, the resolution advisory is released, and the radar drone is commanded to resume its original flight plan. Figure 12 and Figure 13 show this maneuver.

RESOLUTION ADVISORY EXPIRED

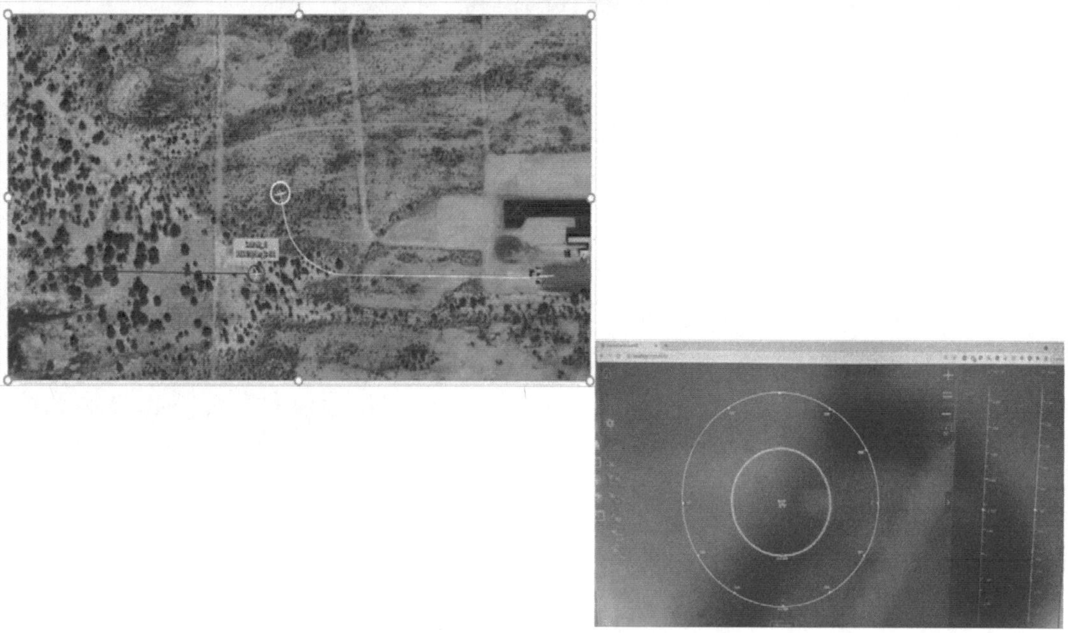

Figure 12, Resolution Advisory Expired, Radar Drone returns to original course.

RADAR DRONE RESUMES ORIGINAL FLIGHT PATH

Figure 13, Radar Drone Resumes Original Flight Path

CONCLUSIONS:

1- The result of this Honeywell Detect and Avoid flight test campaign proved the technical viability of implementing a fully autonomous, low altitude, full integrated avoidance maneuver in a low C-SWAP package.
2- The Honeywell DAA algorithm is fully adaptable and compliant to ACAS-sXu requirements.
3- Optimization of the correlator algorithm is crucial to the Detect and Avoid algorithm this prevent the DAA algorithm to not be overwhelmed with invalid detections and to only use valid target detections for the auto guidance maneuver.
4- The proven ability of the RDR-84K to act as a "standalone" Detect and Avoid edge sensor is a powerful market driver ensuring low C-SWAP requirements are met on cost and weight sensitive UAS and AAM platforms.
5- A radar based Detect and Avoid system offers significant advantages over other sensors (LiDAR, camera others) because of its all weather, day / night capability.
6- The RDR-84K is designed to work in conjunction with other sensors ADS-B to bring the added capability of co-operative Detect and Avoid coverage.

ACKNOWLEDGMENTS

APPENDIX A: ADDITIONAL FLIGHT TEST INFORMATION, TEST CARDS AND TEST PATTERNS

Within each of the flight paths the nine altitude and speed sub-paths were defined tables in this Appendix show each of the flight path definitions. To ensure the highest level of safety prior to flying each of the individual tests a reconnaissance flight was conducted to ensure flight path and navigation alignment to test point M was assured.

Based a-priori information (obtained from rehearsal fights) the helicopter crew would call out (via VHF radio communications) time to the 100 m "no transgression zone" (in this case defined on the test cards as "RGAP", "VGAP" and "LGAP") for each of the test speeds. These time call outs would then trigger the drone pilot to start the radar drone inbound run to Point M, allow the drone to come up to the required 20 m/s and coordinate with the intruder helicopter to the radar convergence area. At the same time the ground station radar controller would start recording radar data until the convergence points where the test was concluded, and helicopter proceeded to its prescribed outbound leg.

In addition to the test cards, Honeywell conducted several pre-planned reconnaissance flights to plot individual way points for each of the flight profiles outlined in the test cards. Figure 3 6 through Figure 3 8 below show those way point paths for each of the flight test patterns.

The magenta vector line shown in each test pattern was used by the helicopter to align to the required altitude and speed called for in each individual test scenario of the test card. Using this method, the helicopter flew this outbound (magenta) path, to establish altitude and speed for the next scenario then turned to an inbound heading to the final approach to the prescribed no-transgression zone to Point M.

Based real-time radar data plots observed during the test flight, a determination was made to either repeat the test, ground communications to the helicopter to "repeat Romeo X" or "proceed to Romeo Y".

ROMEO TEST CARD AND FLIGHT PROFILE

Card 1 Romeo	SmTO MCP Romeo Pattern, Straight in Rwy 28, R44 from east in right traffic Test Point	Vne	HOGE
__R-Recon E95 el 3831' H-TPA 4331' 272° MFLD	092° >>>Rwy 10 3831' TDZE 3824' Rwy 28 <<< 272" 500'agl Mins **4331'** __0 On the ground, Wx & Set baro, update all Mins targets __1 At Orbit, **Radio Check** __2 Foreflight: Time to RGAP (LAST waypoint), early GA turn RT	**60 kt Gs** 15"mp	**272° MFLD 2.1 nm** Time RGAP
__R-1 GA TO RIGHT	361'agl Mins **4192'**	**39 kt Gs** 15.5"mp 0.1 nm Ø, 30°bk	MFLD 272° **0.8 nm** Time to RGAP GA abeam DC3nose
__R-2 GA TO RIGHT	230'agl Mins **4061'**	"	"
__R-3 GA TO RIGHT	295'agl co-Alt Mins **4126'**	"	"
__R-4 GA TO RIGHT	361'agl Mins **4192'**	**78 kt Gs** 17"mp 0.3 nm Ø, 30°bk	MFLD 272° **1.5 nm** Time to RGAP GA mid-1000' mkrs
__R-5 GA TO RIGHT	230'agl Mins **4061'**	"	"
__R-6 GA TO RIGHT	295'agl co-Alt Mins **4126'**	"	"
__R-7 GA TO RIGHT	361'agl Mins **4192'**	**117 kt Gs** 21.5"mp LIMIT 0.7 nm Ø, 30°bk	MFLD 272° **2.1 nm** Time to RGAP GA helipads-1k' mkrs
__R-8 GA TO RIGHT	230' Mins **4061'**	"	"
__R-9 GA TO RIGHT	295'agl co-Alt Mins **4126'**	"	"
__	Card 1, R pattern complete		

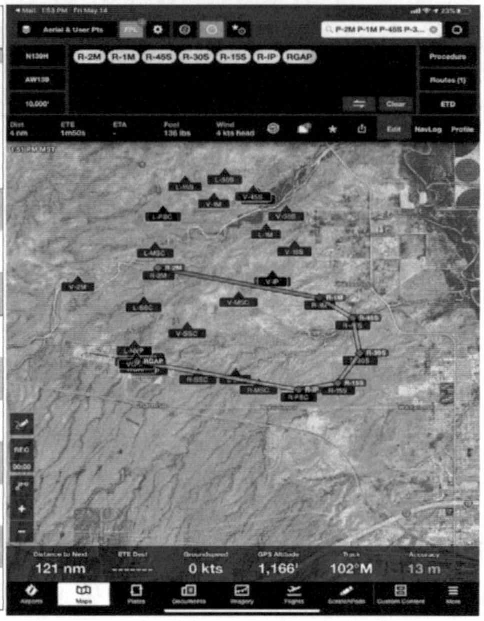

Figure A-1, Romeo Test Cards and Flight Profiles

VICTOR TEST CARD AND FLIGHT PROFILE

Card 2 Victor	SmTO MCP Victor Pattern, R44 from northeast in right turns, RIGHT GA Test Point	Vne	HOGE
__V-Recon E95 el 3831' H-TPA 4331' 227° MFLD	092° >>>Rwy 10 3831' TDZE 3824' Rwy 28 <<< 272° 500'agl Mins **4331'** __0 On the ground, Wx & Set baro, update all Mins targets __1 At Orbit, **Radio Check** __2 Foreflight: Time to VGAP (LAST wpt), early GA turn RIGHT	**60 kt Gs** 15"mp	**227° MFLD 2.1 nm** Time VGAP
__V-1 GA TO RIGHT	361'agl Mins **4192'**	**39 kt Gs** 15.5"mp 0.1 nm Ø, 30°bk	MFLD 227° **0.8 nm** Time to VGAP GA inside e-w rd
__V-2 GA TO RIGHT	230'agl Mins **4061'**	"	"
__V-3 GA TO RIGHT	295'agl co-Alt Mins **4126'**	"	"
__V-4 GA TO RIGHT	361'agl Mins **4192'**	**78 kt Gs** 17"mp 0.3 nm Ø, 30°bk	MFLD 227° **1.5 nm** Time to VGAP GA outside e-w rd
__V-5 GA TO RIGHT	230'agl Mins **4061'**	"	"
__V-6 GA TO RIGHT	295'agl co-Alt Mins **4126'**	"	"
__V-7 GA TO RIGHT	361'agl Mins **4192'**	**117 kt Gs** 21.5"mp LIMIT 0.7 nm Ø, 30°bk	MFLD 227° **2.1 nm** Time to VGAP GA at N-S rd
__V-8 GA TO RIGHT	230' Mins **4061'**	"	"
__V-9 GA TO RIGHT	295'agl co-Alt Mins **4126'**	"	"
__	Card 2, V pattern complete		

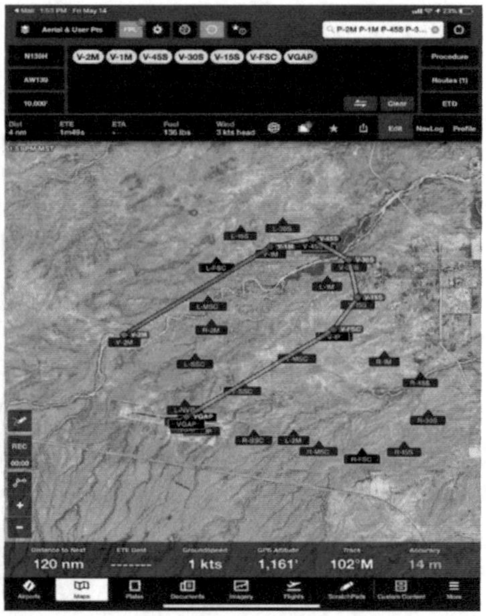

Figure A-2, VICTOR Test Cards and Flight Profiles

LIMA TEST CARD AND FLIGHT PROFILE

Card 3 Lima	5mTO MCT	Vne	HOGE
	Lima Pattern, 90° angle, R44 final from north with GA Left turn Test Point		
__L-Recon E95 el 3831'	092° >>>Rwy 10 3831' TDZE 3824' Rwy 28 <<< 272°		
H-TPA 4331'	500'agl Mins 4331'	60 kt Gs 15"mp	182° MFLD 2.1 nm Time LGAP
182° MFLD	__0 On the ground, Wx & Set baro, update all Mins targets __1 At Orbit, Radio Check __2 Foreflight: Time to LGAP (LAST wpt), early GA turn LEFT		
__L-1 GA TO LEFT	361'agl Mins 4192'	39 kt Gs 15.5"mp 0.1 nm Ø, 30°bk	MFLD 182° 0.8 nm Time to LGAP Left GA at LGAP
__L-2 GA TO LEFT	230'agl Mins 4061'	"	"
__L-3 GA TO LEFT	295'agl co-Alt Mins 4126'	"	"
__L-4 GA TO LEFT	361'agl Mins 4192'	78 kt Gs 17"mp 0.3 nm Ø, 30°bk	MFLD 182° 1.5 nm Time to LGAP GA by LGAP
__L-5 GA TO LEFT	230'agl Mins 4061'	"	"
__L-6 GA TO LEFT	295'agl co-Alt Mins 4126'	"	"
__L-7 GA TO LEFT	361'agl Mins 4192'	117 kt Gs 21.5"mp LIMIT 0.7 nm Ø, 30°bk	MFLD 182° 2.1 nm Time to LGAP GA prior LGAP
__L-8 GA TO LEFT	230' Mins 4061'	"	"
__L-9 GA TO LEFT	295'agl co-Alt Mins 4126'	"	"
__	Card 3, L pattern complete		

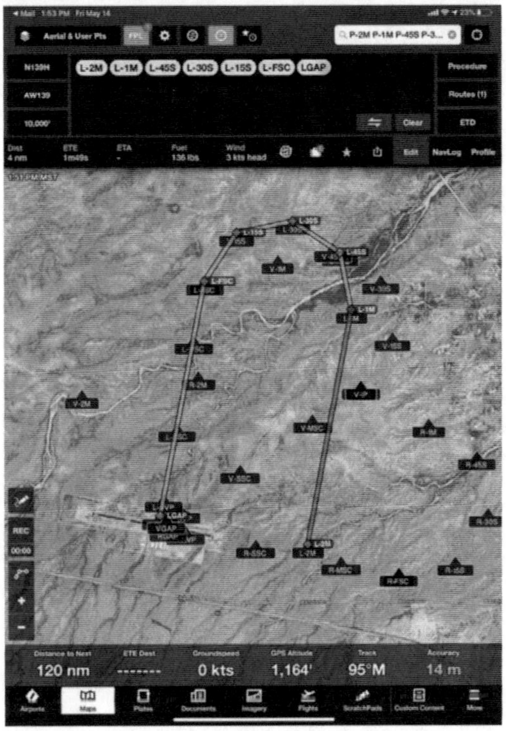

Figure A-3, LIMA Test Cards and Flight Profiles

APPENDIX B: RADAR DRONE FLIGHT CHARACTERISTICS

The BFD mission planner software was used to ensure repeatability in both speed and altitude Figures B-1 and B-2 are overlay plots of multiple day's radar vehicle altitude and speed flight profile showing an altitude delta of less than .2-meter deviation and minor (negligible) deltas of vehicle speed. At the conclusion of each day's testing similar plots were generated to ensure altitude and speed repeatability and report any variation, none were found during the entire flight test.

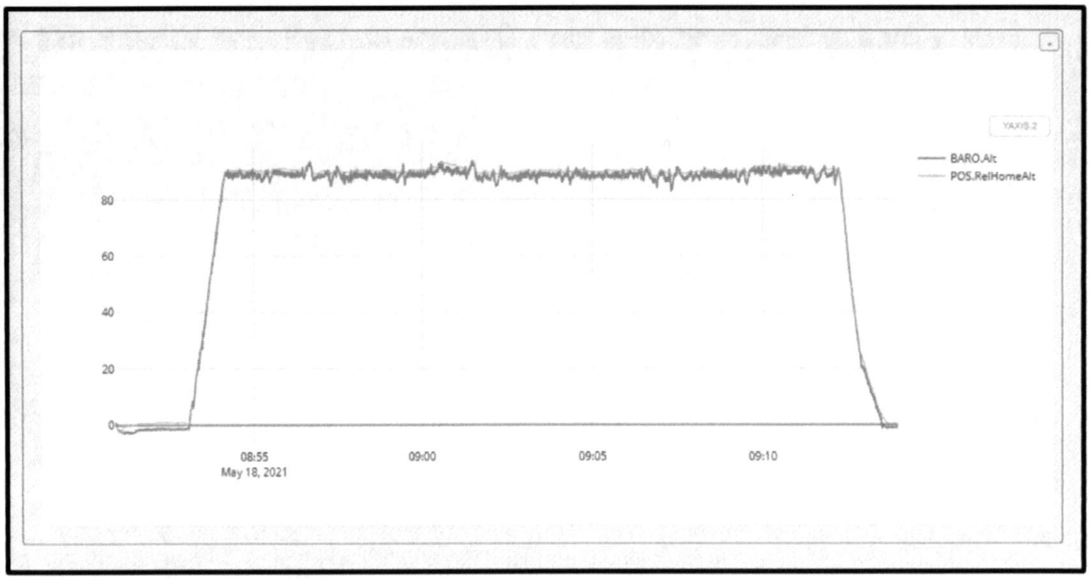

Figure B-1, Radar platform altitude deviation graph

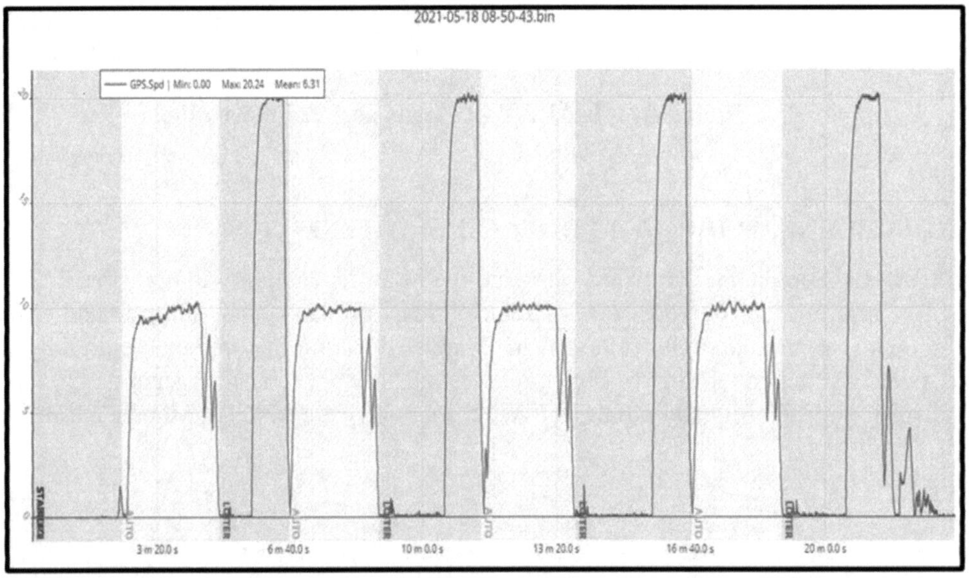

Figure B-2, Radar platform speed deviation graph

THE EFFECTS OF REMOTELY PILOTED AIRCRAFT CONTROL LATENCY DURING AIR-TO-AIR COMBAT

David L. Thirtyacre*

The type of military missions conducted by remotely piloted aircraft continues to expand into all facets of operations including air-to-air combat. While future within-visual-range air-to-air combat will be piloted by artificial intelligence, remotely piloted aircraft will likely first see combat. The purpose of this study was to quantify the effect of latency on one-versus-one, within-visual-range air-to-air combat success during both high-speed and low-speed engagements. The research employed a repeated-measures experimental design to test the various hypothesis associated with command and control latency. Participants experienced in air- to-air combat were subjected to various latency inputs during one-versus-one simulated combat using a virtual-reality simulator and scored on the combat success of each engagement. This research was pursued in coordination with the Air Force Research Laboratory and the United States Air Force Warfare Center.

INTRODUCTION

The military use of remotely piloted aircraft (RPA) continues to increase into principal facets of military aviation. The MQ-1 Predator and MQ-9 Reaper have proven the utility of unmanned aircraft systems (UAS) in combat and amassed millions of flight hours (USAF, 2015a; USAF, 2015b). Since the first introduction of the MQ-1 to the Bosnian theater of operation in 1995, the main mission of the medium-altitude, long-endurance (MALE) RPAs is intelligence collection and ground attack (Trsek, 2007). In the Department of Defense (DoD) mission taxonomy, this includes Intelligence, Surveillance, and Reconnaissance (ISR), and Close Air Support (CAS; Department of Defense, 2014). Despite not being designed or tasked for air-to-air combat, U.S. RPAs have engaged in air-to-air combat, albeit on a very limited scale (Gertler, 2012). An inherent drawback of RPA is the latency of command and control (C2) transmissions (Trsek, 2007). While latency influences all teleoperations, the extent of the effect during within-visual-range (WVR) air-to-air combat has not been explored. In preparation for the use of RPAs and artificial intelligence (AI) in WVR combat, the effects of latency or missing sensor data during a dogfight must be understood. The purpose of this research was to quantify the effect of latency on one-versus-one, within-visual-range air-to-air combat success during both high-speed and low-speed engagements.

Since the advent of the fighter plane in WW-I, every western trained fighter pilot has learned the three axioms of air-to-air combat, lose sight lose fight, maneuver in relation to the bandit, and energy versus nose position (Boelcke, 1916; United States Navy [USN], 2016). These three central themes permeate visual air-to-air combat tactics and describe the importance of analyzing the adversary's current position and state, executing offensive and defensive maneuvers based on the bandit's plane of motion, and making continuous decisions about conserving or exploiting energy (United States Air Force [USAF], 2005). The common thread in these concepts is time. Losing sight of the bandit momentarily, maneuvering too early or late, or depleting energy at the wrong time, all spell defeat in the dogfight (USN, 2016). Boyd (1977) codified these ideas in his Observe-

* Chair, Department of Flight, Worldwide Campus, Embry-Riddle Aeronautical University

Orient-Decide-Act (OODA) Loop Theory and described the desire to complete the loop faster than the adversary in order to dominate air-to-air combat.

Air-to-air combat typically requires a fighter aircraft that is highly maneuverable, capable of transonic velocities, and able to sustain high acceleration loads (Mayer, 2015; Trsek, 2007). These attributes are especially important during WVR combat where two aircraft are entangled in a rapidly changing, highly dynamic fight, with each attempting to gain an advantage and employ ordnance (Shaw, 1985; USAF 2005). While there are reports of short skirmishes between U.S. RPAs and manned enemy fighters, U.S. RPAs were not suited for such an engagement and ultimately defeated (Gertler, 2012). Current MALE UAS do not possess the attributes required to succeed in this dynamic air combat environment (Gertler, 2012; Mayer 2015). However, advances in UAS technology will inevitably yield an aircraft suited for WVR combat (Mayer, 2015; Shin, Lee, Kim, & Shim, 2018; USAF, 2009). As these fighter-unmanned combat aerial vehicles (F-UCAV) become operational, the opportunity for a WVR engagement increases. The first of these engagements will likely be between an F-UCAV and a manned fighter aircraft in an area of responsibility (AOR) far from the ground control station (GCS). Trsek (2007) identified command and control delay as a major hurdle in F-UCAV air-to-air combat and concluded that "…it is presumptuous to assume that short-range engagements are a thing of the past" (p. 26).

The majority of combat missions employing MALE UAS occur thousands of miles from the GCS using terrestrial and satellite communications architecture (Zhang, Fricke, & Holzapfel, 2016). During these beyond-line-of-sight (BLOS) operations, the C2 signal from the GCS must travel through terrestrial networks, be uplinked to a satellite constellation, and then downlinked to the UAS. Telemetry data and sensor information travel the same path in reverse before it reaches the pilot in the GCS. This communication pathway injects latency between the adversary's true position and what is displayed to the pilot. This same latency occurs between the pilot's input and the aircraft receiving the command. Typically, in BLOS operations, the one-way latency can be as low as 0.25 seconds and as high as 1.0 seconds (de Vries, 2005; B. Opp, personal communication, May 15, 2018). During completely autonomous AI operations, delayed, inaccurate, and jammed sensors will influence the fight similar to C2 latency.

The latency can be applied to Boyd's (1977) OODA Loop as delays in observing, difficulty orienting, latent decisions, and delaying the act phase. The delay between the transmitted video/telemetry of the RPA and its reaching the pilot corresponds to the Observe phase and the delay between the RPA pilot making a flight control input and the RPA receiving the command corresponds to the Act phase (remote manipulation) in the OODA Loop. The sum of these two latencies is the total feedback loop latency induced by command and control transmission. However, the effect of transmission latency while maneuvering against a changing target location adds another level of complexity.

The review of relevant literature reveals a distinct gap: the effect of latency during highly dynamic maneuvering while both the vehicle and objective are rapidly changing parameters. This literature gap aligns with Boyd's (1977) OODA Loop Theory, forms the theoretical construct of this study, and defines the independent variables.

The research questions were focused on the effects of latency while executing the phases of Byod's (1977) OODA Loop Theory and comparing the results between high-speed and low-speed engagement entry conditions. The study focused on the control loop latency (input to feedback) in order to isolate the effects. The latency input through IV_1 can be seen as the delay from control manipulation to the aircraft movement plus the return delay.

RQ1: To what extent do different levels of command and control latency affect combat success during one-versus-one, WVR, air-to-air combat?

RQ2: To what extent does initial engagement speed affect combat success during one-versus-one, WVR, air-to-air combat?

RQ3: What is the possible interaction between command and control latency and engagement speed during one-versus-one, WVR, air-to-air combat?

EXPERIMENTAL METHOD

This quantitative research employed a repeated measures experimental design during air-to-air combat simulation. The design allowed multiple, randomized, single-blind treatments of each subject, including a no-treatment control measurement (Verma, 2016). Each subject experienced all of the treatments (6) for each type of engagement (high-speed and low-speed) assigned in the order specified through a balanced Latin square during a 1.0 hour simulation session.

Population/Sample

All fighter pilots are trained in air-to-air combat, but the level of training and proficiency can vary depending on the aircraft and mission. In order to ensure tactical currency and maintain a homogenous population, participants were current fighter pilots who have maintained flight currency in the past five years. All participants have completed basic and advanced air-to-air training and have achieved a qualification equivalent to 4-Ship Flight Lead (USAF) or Division Lead (USN and USMC). Only manned fighter pilots with air-to-air mission qualification in aircraft such as the F-15C, F-15E, F-16C, F-18, F-22, and F-35 were considered. Pilots who have graduated from USN Top Gun or USAF Weapons Instructor Course were preferred due to their advanced knowledge, training, and proficiency.

The sampling strategy purposefully selected participants from the sampling frame. The principal investigator initiated selection ensured purposeful sampling was maintained (i.e., ensuring a mix of pilots from different fighter aircraft). Participant prescreening, management, and scheduling was accomplished through face-to-face discussions to ensure each subject met the participant qualifying criteria. This information was destroyed after the completion of the simulation to ensure confidentiality of the participants. Authorization to collect data was obtained through the institutional review board process at Embry-Riddle Aeronautical University and the Air Force Research Laboratory.

Simulation

The experiment took place in a purpose-built simulator funded through a grant from Embry-Riddle Aeronautical University (ERAU). A unique requirement for this experiment was inducing a system delay. To achieve this, original computer code was developed in the form of a unique Windows driver for which the PI retains copyright. The driver is delay-selectable allowing an input range from 0.000 to 2.000 seconds in 0.001 second increments. The delay was placed between the pilot controls and the simulation software allowing the PI to manipulate IV_1

The Independent Variables (IV), often referred to as the within-subjects factors, are the total, round-trip latency (IV_1) induced into the simulation system through the delay driver and the engagement type (IV_2). The IV_1 is operationalized by assigning the given latency to the delay driver. The second independent variable (IV_2) is the engagement entry speed: high-speed or low-speed. The specific engagement type was operationalized by the engagement starting parameters. The subjects experienced each engagement type six times with the corresponding treatment of IV_1 varying on each test run. Therefore, each subject completed 12 test runs during the simulation.

The Dependent Variable (DV) is the calculated combat score of the engagement. The score was derived from specific angles at the conclusion of the engagement as defined by Shin, Lee, Kim, and Shim (2018) and described in USAF (2005). The CS is defined by Equation (1) where λA is the deviation angle (in radians) of the attacker and λT is the deviation of the target as defined in Figure 1.

$$CS = \frac{(\pi - |\lambda A|)^2 - (\pi - |\lambda T|)^2}{\pi^2} \tag{1}$$

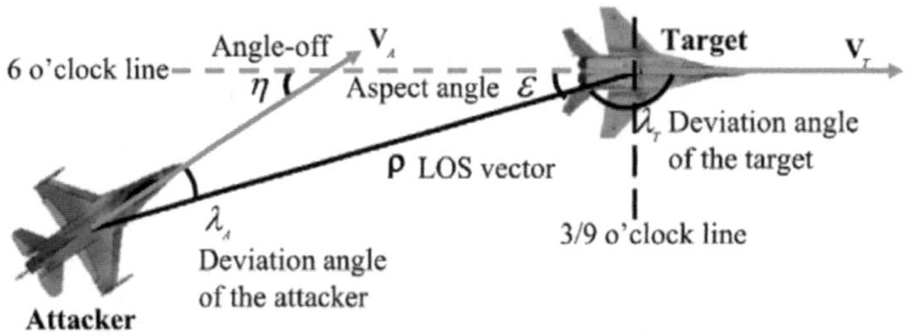

Figure 1. Angular definitions for computation of CS.

Adapted from "An autonomous aerial combat framework for two-on-two engagements based on basic fighter maneuvers" by H. Shin, J. Lee, H. Kim, and D. Shim, 2018, *Aerospace Science and Technology, 72,* p. 306. Copyright 2017 by Elsevier Masson SAS.

While the computation of CS does not directly measure combat success, it codifies the likely outcome of the engagement as described by USAF (2005) and Shin et al. (2018). The CS is, in effect, the normalization of a geometric relationship between the attacker and the target where 1.0 equates to the optimal offensive position (i.e., attacker directly behind and pointing at the target where λA = 0 and λT = π) and -1.0 indicates the worst possible defensive position (i.e., attacker directly in front of the target where λA = π and λT = 0).

Data Collection Process

The simulation period began with a 5-minute introduction including the purpose of the research, description of the simulation, the subject's objective during simulation, safety, and acknowledgment of pertinent IRB and consent information. Following the scripted introduction, each subject experienced a 10-minute simulation familiarization and practice period followed by 12, 2-minute recorded experimental data runs (i.e., 6 runs for high-speed and 6 runs for low-speed). The objective of the familiarization and practice period was for the subject to become accustomed to the performance of the simulated aircraft, become familiar with the simulation environment, and comfortable with the VR headset. While many repeated measure experiments attempt to avoid learning or the carry-over effect, this experiment intended for the subjects to learn as much as possible before data collection (Gorsich et al., 2018). The practice period also aligns with the fact that all military pilots receive extensive training on their weapon systems to understand its strengths and weaknesses before entering combat. The familiarization and practice period also reduced the effects of confounding variables such as non-familiarity with the simulation hardware/software, control manipulation, and display systems.

The experimental sequence consisted of 12 engagements with an approximate duration of 120 seconds each. Based on results from the field test, the high-speed engagement concluded after 105 seconds while the low-speed engagement concluded in 90 seconds. After each engagement, there was a 45-60 second rest period prior to the next run. For each engagement, one of the six preset latency categories was assigned through a balanced Latin square design until all latency levels were experienced by each subject, on each engagement type.

The parameters of each engagement were closely controlled. The data runs for each category of fight (i.e., high-speed and low-speed) began from the same starting point, altitude, and range saved in the primary test profile. However, each engagement varied the adversary starting velocity vector introducing slight differences in the engagement geometry; this input was made to decrease predictability. Both the target and the attacking aircraft remained the same (airframe performance, visual depiction, and avionics). The high-speed simulation runs began with the attacker (subject) placed 3.5 NM from the target aircraft while the low-speed engagements were from a line-abreast formation. For the high-speed engagement, both aircraft began the engagement at 450 KTAS (232 m/s), 20,000 feet (6,098 m) mean sea level (MSL) and approximately pointed at the other aircraft as defined by USAF (2005) and USN (2016). Each low-speed simulation run began with both aircraft at 250 KTAS (129 m/s), in line abreast formation, heading in the same direction. These parameters are similar to those stated in USAF (2005) and USN (2016) as typical high aspect WVR starting parameters. The adversary (target) flight AI profile was set to "expert" commanding the target aircraft to attempt to shoot the attacker throughout the engagement.

Each engagement concluded at a time specified by the field test. Since an engagement CS changes throughout the fight, angles (and CS) were assessed at multiple times during the engagement. The assessment occurred near the end of the engagement and consist of three measurements which were at start + 1:15, 1:30, and 1:45 for the high-speed engagements and start + 1:00, 1:15, and 1:30 for the low-speed engagements. The assessment times were determined during the field test. All engagements were recorded through the simulation system at a parametric update rate of greater than 10Hz for posttest analysis and data collection.

A two-way repeated measures ANOVA tested whether the mean of the dependent variable varied by latency treatments levels and each tactical scenario (i.e., high-speed and low-speed). The results determined if there was a significant difference between the treatments on the population and if there was an interaction effect between the IVs. If there is a significant statistical difference ($p < .05$), then further analysis was performed with custom contrasts (Field, 2013; Verma, 2016). However, the ANOVA only determines if there was a statistical difference in treatments on the DV, not which treatment is significant or how many of the treatments are significantly different. In order to determine which specific treatment is significant, more tests were required. For this, pairwise comparisons are more appropriate than a post hoc test because the DV is expected to decrease with an increased latency and only two treatment levels were compared at once (e.g., latency A compared to latency B).

RESULTS

A sample of 29 participants, which included 348 separate and distinct engagements over the 12 IV combinations, was collected at a military base between October 26, 2020 and November 2, 2020. One participant's data set was removed for outliers. All of the pilots were screened prior to simulation entry to ensure compliance with the sample requirements and confirmed by local personnel. The sample was purposely recruited by word of mouth and included pilots from the F-15C, F-15E, F-16C, F-22, and F35. Simulation periods were both scheduled and ad hoc, and took place in a private office supplied by the base. Due to the active-duty status of subjects involved, the confidential nature of the IRB, and subsequent AFRL approval, no demographics or personally identifiable information was collected. All of the participants were highly experienced fighter pilots.

Each engagement was scored at three assessment times in order to record a reliable measurement of each engagement. The average of the three assessments was used as the combat score for the engagement. The descriptive statistics for all of the qualified engagements are listed in Table 1.

Table 1. Descriptive Statistics of Combat Scores

IV_2	IV_1	N	Mean	SD	Min.	Max.
High-speed	0.000	28	.539	.155	.096	.869
	0.250	28	.491	.152	.098	.821
	0.500	28	.387	.216	-.224	.717
	0.750	28	.314	.192	-.179	.605
	1.000	28	.342	.163	-.061	.790
	1.250	28	.133	.221	-.311	.556
Low-speed	0.000	28	.659	.102	.510	.917
	0.250	28	.616	.121	.314	.932
	0.500	28	.568	.132	.332	.907
	0.750	28	.453	.136	.150	.739
	1.000	28	.308	.200	-.144	.627
	1.250	28	.189	.229	-.469	.668

The mean combat scores for each latency level are plotted in Figure 2 for the high-speed engagements and Figure 3 for the low-speed engagements. The combination of these two charts illustrate the overall mean combat score at each latency level. Figure 4 depicts how CS varied with latency when combining the high and low-speed engagement results. However, this chart should be interpreted with care since there was a statistically significant interaction effect indicating that the latency effect on CS during the high-speed engagements was not the same as during the low-speed engagements.

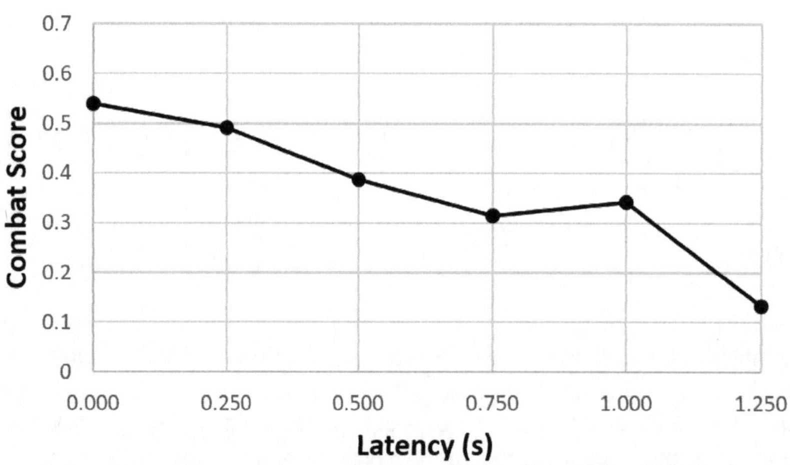

Figure 2. Mean Combat Scores for High-speed Engagements by Latency.

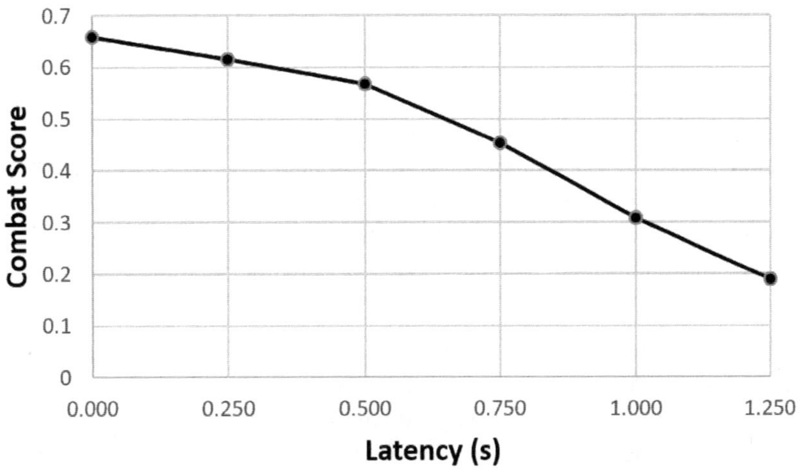

Figure 3. Mean Combat Scores for Low-speed Engagements by Latency.

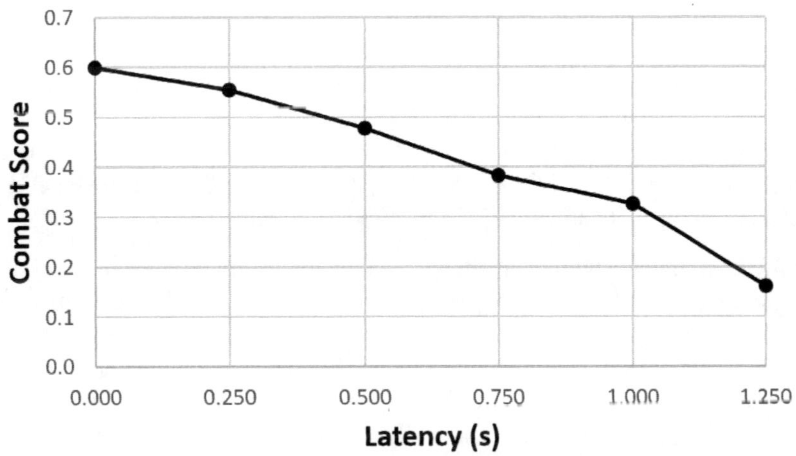

Figure 4. Mean Combat Scores for all Engagements by Latency.

The appropriate assumption testing was completed for the two-way repeated measured design including testing for outliers, sphericity, and normality. The two-way repeated measures ANOVA indicated that there was a significant two-way interaction between engagement speed and latency, $F(5, 135) = 3.71$, $p = .004$, partial $\eta 2 = .121$. This significant interaction indicates that the effect of latency on combat score is dependent on both the amount of latency and the speed of the engagement.

Main Effects

The main effects for a two-way repeated measures ANOVA that has a significant interaction effect must be considered cautiously. Due to the significant interaction, the main effects could be misleading or incomplete. However, exploring the main effects can yield more insight to the relationships. The main effect of engagement speed showed a significant difference between high and low-speeds with a mean difference = .098, $F(1, 27) = 15.62$, $p = .001$, partial $\eta 2 = .367$. While the interaction effect met the assumption of sphericity, the main effect of latency did not $x^2(14) = 31.54$, p = .005. After applying the Greenhouse-Geisser correction, the main effect of latency was also significant $F(3.56, 96.24) = 72.57$, $p < .001$, partial $\eta^2 = .729$.

Simple Main Effects

Since there was a significant interaction effect for the two-way repeated measures ANOVA, analyzing the main effects could yield inaccurate conclusions as to the significance of the findings. Instead, the simple main effects were examined to further understand the relationships between the variables. This was accomplished through use of the one-way repeated measure ANOVA. The test for sphericity was required during analysis of the simple main effects for latency since there are six treatments, but not for engagement speed since there are only two treatments.

The latency during the low-speed engagement met the assumption of sphericity, x2(14) = 21.07, p = .101. The difference in CS for latency treatments during the slow-speed engagements were statistically significant $F(5, 135) = 62.87$, $p < .001$. The latency during the high-speed engagement did not meet the sphericity assumption, x2(14) = 48.63, $p < .001$. The degrees of freedom were corrected using the Greenhouse-Geisser adjustment to correct the validity of the F value. The CS for the latency treatments during the high-speed engagements were statistically significant $F(3.34, 90.27) = 21.69$, $p < .001$. The results of the pairwise comparison for latency during the high-speed engagements are illustrated in Table 2.

Hypothesis Testing Results

The first alternate hypothesis stated that there is a significant decrease in combat success between fighter pilots experiencing no latency and those experiencing latency. Since there was a significant interaction between within-subjects factors, this hypothesis was tested by analyzing the simple main effects of latency. The results indicate that there was a statistically significant decrease in combat score during both high-speed, $F(3.34, 90.27) = 21.69$, $p < .001$, and low-speed engagements, $F(5, 135) = 62.87$, $p < .001$. Therefore, the null hypothesis was rejected.

The second alternate hypothesis stated that there is a significant difference in combat success based on initial engagement speed. Since there was a significant interaction between within-subjects factors, this hypothesis was tested by analyzing the simple main effects of engagement speed on combat score at each latency level. The results indicate that there was a statistically significant difference in combat score between high-speed and low-speed engagements with latency levels of 0.00, 0.25, 0.50, and 0.75 seconds. Therefore, the null hypothesis was rejected for these latency levels. However, for latency levels of 1.00 and 1.25 seconds, there was no statistically significant difference between engagements fought at high-speed versus low-speed. Therefore, the findings fail to reject the null hypothesis for those latencies.

The third alternative hypothesis stated that there is a significant interaction between command and control latency and engagement speed. This hypothesis was tested with the two-way repeated measures ANOVA which indicated there was a significant interaction effect between factors on the dependent variable, $F(5, 135) = 3.71$, $p = .004$, partial $\eta2 = .121$. Therefore the null hypothesis was rejected.

Table 2. Pairwise Comparison

Pairwise Comparison between Latencies by Engagement Speed (Simple Main Effects)

	High-speed Engagement				Low-speed Engagement		
Latency (s)	Latency(s)	Mean Diff	Sig.	Latency(s)	Latency(s)	Mean Diff	Sig.
0.000	0.250	.047	1.000	0.000	0.250	.043	1.000
	0.500	.152	.161		0.500	.091*	.038
	0.750	.225*	.002		0.750	.206*	.000
	1.000	.197*	.000		1.000	.351*	.000
	1.250	.406*	.000		1.250	.470*	.000
0.250	0.000	-.047	1.000	0.250	0.000	-.043	1.000
	0.500	.105	.230		0.500	.048	.955
	0.750	.177*	.023		0.750	.163*	.000
	1.000	.149*	.002		1.000	.308*	.000
	1.250	.358*	.000		1.250	.427*	.000
0.500	0.000	-.152	.161	0.500	0.000	-.091*	.038
	0.250	-.105	.230		0.250	-.048	.955
	0.750	.073	1.000		0.750	.115*	.002
	1.000	.045	1.000		1.000	.260*	.000
	1.250	.254*	.000		1.250	.380*	.000
0.750	0.000	-.225*	.002	0.750	0.000	-.206*	.000
	0.250	.177*	.023		0.250	-.163*	.000
	0.500	-.073	1.000		0.500	-.115*	.002
	1.000	-.028	1.000		1.000	.145*	.002
	1.250	.181*	.001		1.250	.265*	.000
1.000	0.000	-.197*	.000	1.000	0.000	-.351*	.000
	0.250	-.149*	.002		0.250	-.308*	.000
	0.500	-.045	1.000		0.500	-.260*	.000
	0.750	.028	1.000		0.750	-.145*	.002
	1.250	.209*	.001		1.250	.120	.195
1.250	0.000	-.406*	.000	1.250	0.000	-.470*	.000
	0.250	-.358*	.000		0.250	-.427*	.000
	0.500	-.254*	.000		0.500	-.380*	.000
	0.750	-.181*	.001		0.750	-.265*	.000
	1.000	-.209*	.001		1.000	-.120	.195

* indicates a statistically significant difference in mean combat score.
Note: Values of 1.000 are an artifact of SPSS when using the Benferroni adjustment and indicate a non-significant relationship, not an actual value of 1.000. This is to avoid an alpha error accumulation through multiple pairwise comparisons.

DISCUSSION

The results of this experiment clearly illustrate the effect of latency and engagement speed on combat success during a WVR fight. However, there are several areas worthy of further examination including the performance of the simulated aircraft and the theoretical and practical implications of the research. However, before discussing the conclusions of this study, it is important to consider the performance of the simulated aircraft and adversary aircraft.

While the results of this study indicate that pilots can still gain, and maintain, an offensive position even at the highest tested latency, consideration must be given to the superior performance of the simulated aircraft. During the experimental runs, subjects often max-performed the aircraft resulting in acceleration loads as high as 11.0 Gz while the maximum observed adversary load was 7.3 Gz. This was especially true at higher latency levels when the pilots found themselves in poor tactical position and used superior aircraft performance to outmaneuver the adversary. There was a similar observation for the aircraft angle of attack. While the maximum observed AoA for the adversary was 25.2 degrees, the subjects routinely maneuvered the simulated aircraft to AoA greater than 35 degrees (indicated by a warning tone) and in some instances, as high as 56 degrees.

It was clear that the superior performance of the simulated aircraft influenced the combat outcome of the engagements. However, this was an intentional aspect of the test plan designed to give pilots a maneuvering advantage similar to what would be available in an F-UCAV (Trsek, 2007). While the specific combat score was undoubtedly influenced by aircraft performance, it was clear that the decrease in performance is present regardless of the superior performance of the F-UCAV. Therefore, the conclusions of this study should be taken as degradation of combat effectiveness (i.e., the difference between engagements without latency and those with latency) and not a specific value of combat success.

Finally, the data, observation, and engagement playback lead to the conclusion that there were several effects of latency of which the pilot must contend including lift vector control, airspeed control, and general aircraft control. At lower latencies, the main obstacle was lift vector orientation and control. While the pilots may know where the optimal location of their lift vector should be, the latency caused them to either undershoot or overshoot the desired position (i.e., rolled past the desired position). As latency increased, this issue was compounded, often leading to an orientation in the opposite direction than desired. Latencies of 0.750 seconds and above contributed to large variations in airspeed since the throttle and speedbrakes were also delayed as part of the command and control link. These large energy excursions led to a larger than desired turn radius or a lack of energy required to complete a maneuver. The airspeed control issues coupled with poor lift vector control often resulted in difficulty controlling the aircraft. These results are illustrated in Figure 5.

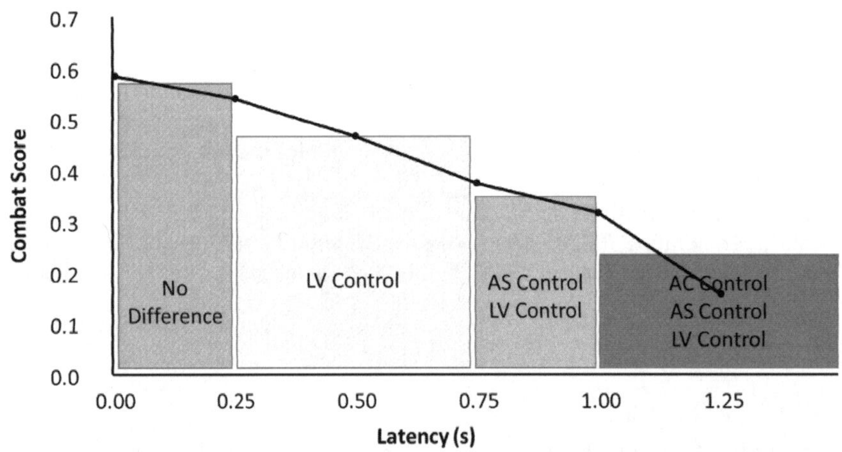

Figure 5. Observed piloting issues associated with latency.
Note: Mean Combat Score values displayed are based on the main effects of both engagement speeds

CONCLUSION

The theoretical foundation of this study was the OODA Loop (Boyd, 1977). While the original construct of the OODA Loop theory was based around making tactical decisions faster than the adversary, this study indicates that technology-based latency influences the engagement outcome similar to a slow decision-making cycle. This is foundation to the understanding of the OODA Loop since, in its original form, it described the human decision-making process where the individual observes an action, orients based on knowledge and previous experience, decides on an action, and executes the action. This study adds depth to the theory illustrating that technology-induced latency has a similar effect as slow human decision making resulting in lower performance. Therefore, latency, when combined with the human decision-making process, compounds the effect resulting in significantly lower performance.

The current understanding of the OODA Loop process was that command and control latency would only affect the Observe and Act phases of the OODA Loop. However, this study indicates that latency affects the entire OODA Loop and that the Orient-Decide-Act process was particularly influenced by latency. The ability of the pilots to maintain congruency between orientation and action proved more difficult as latency increased. This caused the pilots to spend most of their time in the O-D-A phases while occasionally returning to the Observe phase. An accurate analogy would be that the pilots were stuck in a Do-Until Loop between orientation, decision, and action as illustrated in Figure 6. The Do-Until loop was continued until the action determined in the decide phase was satisfactorily completed. Other studies of latency identified the move-and-wait strategy to compensate for delays in command and control; the effect seen in this study could be interpreted as a dynamic move-and-wait (Marano et al., 2015; Sheridan & Ferrell, 1963; Storms et al., 2017; Vozar et al., 2017).

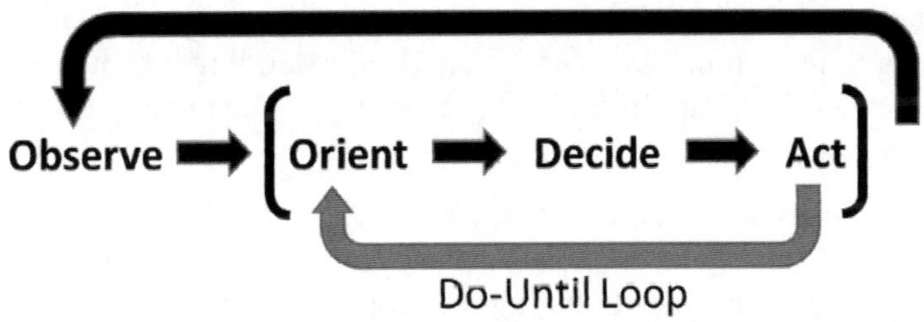

Figure 6. Illustration of Do-Until loop acting internal to OODA Loop process.

There are several practical outcomes of the study that are of particular interest. Although there was a significant decrease in combat score with increased latency, pilots were able to maintain an offensive advantage even at the highest tested latency. As mentioned above, this could be partially attributed to the superior performance of the simulated aircraft, but also supports the conclusion that given enough performance advantage, an offensive position is possible even with a 1.250 second latency.

The results of the Field Test effectively bounded the upper limit of latency based on manual aircraft control. When latencies of 1.500 seconds and above were tested, severe aircraft control issues emerged, often resulting in ground impact during engagements. At the other end, the experimental results revealed that a latency of 0.250 seconds was not significantly different than the combat scores without latency. These results support the conclusion that C2 latencies of 0.250 seconds and below are acceptable and that latencies above 1.250 seconds are unacceptable for a manually controlled aircraft. The results also support the conclusion that latencies greater than 0.250 seconds, but less than 1.250 seconds may be at least partially offset by superior aircraft performance during both high-speed, two circle engagements and low-speed, one-circle engagements (USAF 2005; USN 2016).

The experimental results indicate that there was not a significant difference in CS between zero latency and 0.500 seconds of latency during the high-speed, two circle fight. Additionally, there was not a significant difference between 0.500 and 1.000 seconds of latency for the high-speed fight. A possible conclusion stemming from these results is that the two-circle fight is less susceptible to degradation due to latency. This conclusion is supported by observation during the experiment that orientation and maneuvering was easier during the two-circle fight versus the one-circle fight where the lift vector orientation changes rapidly. The practical application of these results is that when latency above 0.250 seconds is present, the two-circle fight is desired over the low speed one-circle fight.

In a few cases, subjects achieved very high combat scores even at the highest tested latency. One subject was able to achieve an average engagement score of .668 with a latency of 1.250 seconds. Results like this indicate that pilot technique may play a larger role than expected in countering the effects of latency and should be explored in future studies.

A final and ancillary practical contribution of this study was demonstrating that a properly configured VR simulator can produce an effective air-to-air training environment. While not the purpose of this experiment, the simulation provided an effective and efficient environment in which to practice manual flight skills. This was supported by pilot comments during the experiment, SMEs, and other simulation and aviation experts. Considering that the cost of the VR simulation was approximately 1/1000th the cost of a modern fighter simulator, VR simulation should be an integral venue for future training.

It should be emphasized that this study intentionally excluded several variables such as sensors, weapons, weapon cueing, and weapon performance with the objective of isolating the ability of the pilot to maneuver to, and remain in, the control zone. Understanding how latency affects the basic fighter maneuvers employed by the subjects is the first step in developing tactics to overcome latency.

This study has quantified the effects of latency on combat capability during an air-to-air WVR engagement and explored the effect of tactics on the outcome. It has further added depth and breadth to the OODA Loop Theory by including the concept of the dynamic move-and-wait Do-Until loop. It has provided several practical contributions to the problem of command and control latency during air-to-air and outlined several areas for further study.

References

Boelcke, H. (1916). *Dicta Boelcke*. Germany: Die Fliegertruppen des deutschen Kaisserreiches. Retrieved from https://net.lib.byu.edu/estu/wwi/comment/dicta-b.html

Boyd, J. (1977). *Patterns of conflict.* Retrieved from https://danford.net/boyd/patterns.pdf

de Vries, S. C. (2005). *UAVs and control delays.* (TNO report DV3 2005 A054). Netherlands: TNO Defence, Security and Safety. Retrieved from http://www.dtic.mil/dtic/tr/fulltext/u2/a454251.pdf

Department of Defense. (2014). *Joint publication 3-30 command and control of joint air operations.* Washington, DC: U.S. Government Publishing Office. Retrieved from http://www.jcs.mil/Portals/36/Documents/Doctrine/pubs/jp3_30.pdf

Field, A. P. (2013). *Discovering statistics using SPSS: And sex and drugs and rock 'n' roll* (4th ed.). Los Angeles, CA: SAGE.

Gertler, J. (2012). *U.S. unmanned aerial systems* (R42136). Washington, DC: Congressional Research Service. Retrieved from https://www.everycrsreport.com/reports/R42136.html

Gorsich, D. J., Jayakumar, P., Cole, M. P., Crean, C. M., Jain, A., & Ersal, T. (2018). Evaluating mobility vs. latency in unmanned ground vehicles. *Journal of Terramechanics*, 80, 11-19. doi:10.1016/j.jterra.2018.10.001

Marano, A., Priora, F., Melandro, F., & Formisano, G. (2015). History of robotic surgery. In *Robotic Surgery* (pp. 1-12). Milano, Italy: Springer.

Mayer, M. (2015). The new killer drones: Understanding the strategic implications of next-generation unmanned combat aerial vehicles. *International Affairs, 91*(4), 765-780. doi:10.1111/1468-2346.12342

Shaw, R. (1985). *Fighter combat tactics and maneuvering.* Annapolis, MD: Naval Institute.

Sheridan, T. B., & Ferrell, W. R. (1963). Remote manipulative control with transmission delay. *IEEE Transactions on Human Factors in Electronics, HFE, 4*(1), 25-29. doi:10.1109/THFE.1963.231283

Shin, H., Lee, J., Kim, H., & Shim, D. (2018). An autonomous aerial combat framework for two-on-two engagements based on basic fighter maneuvers. *Aerospace Science and Technology, 72*, 305-315. doi:10.1016/j.ast.2017.11.014

Storms, J., Chen, K., & Tilbury, D. (2017). A shared control method for obstacle avoidance with mobile robots and its interaction with communication delay. *The International Journal of Robotics Research, 36*(5-7), 820-839. doi:10.1177/0278364917693690

Trsek, R. (2007). *The last manned fighter: Replacing manned fighters with ucavs.* (Master's thesis). Air Command and Staff College, Maxwell Air Force Base, AL.

United States Air Force. (2005). *Tactics, techniques and procedures 3-3. Washington, DC:* Headquarters, USAF Retrieved from http://falcon.blu3wolf.com/Docs/Basic-Employment-Manual-F-16C-RoKAF.pdf

United States Air Force. (2009). *United States Air Force unmanned aircraft system flight plan 2009-2047. Washington, DC:* Headquarters, USAF. Retrieved from https://fas.org/irp/program/collect/uas_2009.pdf

United States Air Force. (2015a). *MQ-1 Predator* [Fact sheet]. Retrieved from http://www.af.mil/About-Us/Fact-Sheets/Display/Article/104469/mq-1b-predator/

United States Air Force. (2015b). *MQ-9 Reaper* [Fact sheet]. Retrieved from http://www.af.mil/About-Us/Fact-Sheets/Display/Article/104470/mq-9-reaper/

United States Navy. (2016). *Basic fighter maneuvering, section engaged maneuvering, T-45 Strike.* Retrieved from https://www.cnatra.navy.mil/local/docs/pat-pubs/P-1289.pdf

Verma, J. P. (2016). *Repeated measures design for empirical researchers.* Hoboken, NJ: Wiley.

Vozar, S., Storms, J., & Tilbury, D. M. (2016). Development and analysis of an operator steering model for teleoperated mobile robots under constant and variable latencies. *Robotica*, 1-20.

Zhang, F., Fricke, T, & Holzapfel, H. (2016). Integrated control and display augmentation for manual remote flight control in the presence of large latency. *Proceedings from AIAA guidance, navigation, and control conference*. doi: 10.2514/6.2016-1867

BEYOND THE LINE OF SIGHT? WHAT'S NEW IN OPTICAL PERCEPTION

Tristan Swedish[*] [†] and Sebastian Bauer[*]

The technology used in optical perception has made incredible progress in the last few decades, driven by the digital revolution and semiconductor scaling effects. Despite these advances, captured images resemble those of the earliest conception of cameras: the measurement is proportional to the total number of photons arriving at the sensor over a predetermined exposure time. However, emerging single photon capture is an exciting new paradigm for image processing, where single photons are digitized directly when they arrive on the image plane. In this review, we discuss the potential of single photon capture for image processing and transformative applications such as seeing beyond the line of sight.

INTRODUCTION

Practical deployments of autonomous vehicles and mobile robots must be able to operate within a wide range of uncontrolled or unpredictable environments. This lack of predictability can be addressed with sufficiently robust perception, which enables dynamic planning and control. Optical sensing has become an essential part of the autonomous perception stack due to a number of factors. First, autonomous systems often work alongside, or as replacement for humans, and much of our environment is adapted to our biological visual perception system. Second, optical contrast in the visible wavelengths is informative for most natural environments, as evidenced by the convergent evolution of visual systems in nature. Finally, optical imaging systems are a well understood technology with a robust ecosystem of solution providers and academic research interest, driving much of the field of Computer Vision. For these reasons, optical sensing is highly versatile, and can be used to directly perform many perception sub-tasks: depth sensing, localization, object detection, and material classification.

There are a number of challenges for existing perception systems that make use of optical imagery. Environments can in general contain low light, include fast moving objects, and experience dynamic illumination changes requiring high dynamic range of image sensors. Conventional image processing approaches handle each of these problems separately reasonably well, but when more than one are present, the underlying images can contain artifacts such noise, motion blur, or saturation. For example, reducing the exposure time can capture fast objects without motion blur, but this produces extremely noisy images in low light. When a camera is optimized for low light performance, its dynamic range typically suffers as well, or motion artifacts can occur due to

[*] Ubicept, Inc., 55 Court St. Floor 2, Boston, MA
[†] PhD Candidate, MIT Media Lab, Camera Culture Group

combining images under multiple exposures. Fundamentally, any downstream perception task is limited by the quality of its input data.

Conventional Imaging Systems

Conventional imaging systems are typically composed of a conventional image sensor (CIS) and imaging optics. The imaging optics form an image of the scene by refracting light onto a 2D array of light sensitive pixels. Each pixel converts incident photons into electrons, which are stored at each pixel. The number of electrons within each pixel "photon well" can then be estimated by measuring the voltage across a capacitor before being reset. The design of these systems naturally leads to the notion of a "frame." The array of pixels are exposed to light for a given exposure time, and then the analog voltage at each pixel is digitized using an analog to digital converter (ADC). Each exposure produces a new frame and a sequence of frames is used in subsequent processing to enable perception.

There are three primary sources of noise in image frames captured by CMOS sensors: photon shot noise, read noise, and dark current. Photon shot noise is due to the inherently stochastic nature of photon arrival time. Photon flux can only be estimated by the number of discrete photons that arrive within an interval of time, and there always exists some variance due to the underlying Poisson process. Read noise is added by the ADC when reading the accumulated charge at each pixel. Dark current generates additional electrons contributing to the total charge in the photon well. While shot noise is a fact of nature, read noise and dark current are limits of the technology used in CIS.

Photon Count Imaging

Moving beyond the conventional photon integration scheme used in image sensors today, there is an emerging trend to develop *single photon* image sensors. These sensors count individual photons as they arrive at the image sensor. Photon counting is desirable because it couples the measurement scheme with the underlying physics: counting removes sources of noise such as read noise and provides a discrete representation of arriving light that is well suited for digital signal processing. There are two primary types of single photon counting sensors that are currently on the market: Conventional Image Sensor Quanta Image Sensors (CIS-QIS) and Single Photon Avalanche Diode (SPAD) arrays. Both types of sensors can be used with conventional imaging optics and output data from a 2D array of pixels.

Table 1. Image Sensor Comparison between CIS, CIS-QIS, and SPAD arrays. CIS tend to have poor temporal resolution and exposure flexibility due to their relatively long exposure times and read noise. CIS-QIS tends to excel in low light and dynamic range, and is based on the idea of exploiting future high resolution sensors. SPAD arrays have extremely high time resolution, making them useful for light time of flight, but also can be used in passive mode for very high dynamic range. SPADs and CIS-QIS low read noise allows them to flexibly integrate photon detections after capture, making them well suited for computational imaging approaches.

Technology	Spatial Resolution	Temporal Resolution	Low Light	Dynamic Range	Exposure Flexibility
CIS					
CIS-QIS					
SPAD					

CIS-QIS is being commercialized by the company Gigajot, making use of an architecture that is similar to conventional CMOS, but with low read noise and single photo-electron sensitivity. The CIS-QIS pixels, known as jots, produce an analog readout that can be digitized with single photon resolution. CIS-QIS makes use of sub-electron read noise, making it possible to quantize detections to known single photon voltage levels. Low read-noise becomes extremely useful for low-light or photon-starved environments as the small signal can be buried by the sensor read-out used in conventional approaches. CIS-QIS take advantage of the trend in decreasing CMOS pixel size, as the cross-section of small pixels capture less photons, the low-photon count data across blocks of adjacent pixels can be used to estimate a lower variance estimate of photon flux.

SPADs use a much different operating principle from CIS. SPADs use a high-voltage reverse bias photodiode that triggers an avalanche of electrons when a single photon arrives at the sensor. This avalanche produces a signal that can be easily detected using electronics on each pixel. Once an avalanche occurs, the diode must be recharged before making additional detections, producing a small deadtime. A key advantage of SPADs is that the avalanche can be measured with high time precision. Time-Correlated Single Photon Counters (TCSPCs) can be used to precisely time-tag the arrival of photons that are detected by SPADs. For this reason, SPADs have been widely used for lidar (Light Detection And Ranging) applications as the timing resolution is sufficient to produce ranging information using the time-of-flight (ToF) of photons. In lidar, a TCSPC is synchronized with a laser pulse, producing a timing histogram that measures photon counts over a set of delays from the sync signal. These histograms can be used to determine the distance to objects using ToF. SPADs have been used in optics applications for decades, but have typically only been used as single pixel detectors, and have traditionally had high dark count rates (DCR) due to thermal noise. Recently, high resolution SPAD arrays have been introduced and DCR have dropped dramatically. The latest iPhone Pro and iPad[*] models incorporate a SPAD array and vertical-cavity surface emitting laser (VCSEL) array for short-range flash lidar. These improvements to SPADs have motivated their use in passive imaging as well. Canon has announced that they will commercialize a 3.2 MP SPAD array[†] for passive imaging applications in their security camera line of product to be manufactured in FY22.

SINGLE PHOTON PERCEPTION

The emergence of single-photon image sensors has gained interest in the computational imaging community. Principally, the low-level image signal processing required to produce high-quality images differs from CIS. Furthermore, new applications are possible when making use of single-photon detectors. We can approximately distinguish the imaging modes as either *passive* or *active*. Passive imaging utilizes ambient light to form images, such as cameras used for perception in mobile robots. Passive imaging is well suited for general perception in natural environments, and is almost universally employed by the perception system of animals. In the following section, we review some work utilizing single photon imaging that does not require a controlled illumination source.

---

[*] https://www.apple.com/newsroom/2020/03/apple-unveils-new-ipad-pro-with-lidar-scanner-and-trackpad-support-in-ipados/

[†] https://global.canon/en/news/2021/20211215.html

SPAD Image Processing

One of the first works to make use of SPADs for general purpose imaging and highlight their capacity to enable low-light, fast motion imaging was Quanta Burst Photography[1] (QBP). The high level idea of QBP was to integrate single-photon data from high-framerate SPAD arrays in a similar way as multiple frames are combined in the burst photography algorithms used in smartphones. The SwissSPAD2[2] can produce binary frames representing photon detections at ~100k fps, and the non-existent read noise of SPADs makes it possible to integrate over many thousands of binary frames in order to produce sharp, low noise images of dynamic scenes. As in burst photography, QBP outputs high dynamic range (HDR) images that incorporate motion compensation. An alternative to QBP is Motion Adaptive Deblurring[3], which makes use of a changepoint estimation algorithm to adjust the integration time of each pixel independently depending on scene motion.

While integrating photon count data can provide a good estimate of photon flux, Passive Inter-Photon Imaging[4] proposes to measure the time between photon detections to estimate photon flux. The advantage to this approach is that the photon flux can be estimated over an extremely large dynamic range. Since photon counts are generated by a Poisson process, inter-photon timing follows an exponential distribution. Thus, if inter-photon timing is available, the decay rate of this exponential distribution can be estimated with just a few inter-photon timing measurements. Such approaches challenge notions of framerate and latency, and are one example of how SPADs may provide a new paradigm for image processing.

CIS-QIS Image Processing

SPAD-based processing makes use of the high time resolution of SPADs. Similarly, CIS-QIS make use of their high spatial resolution and sensitivity[5]. CIS-QIS produce binary frames at approximately ~1k fps, but at a higher resolution than current SPAD arrays. Similar to SPADs, CIS-QIS has been proposed as input for low-light fast motion imaging[6] and HDR[7]. CIS-QIS have been produced with color filter arrays (CFA), but special considerations must be taken to perform demosaicking[8].

High Level Perception

One of the more recent trends in single-photon perception is to consider using the raw data to perform inference tasks directly, rather than first form an image. Since single photon capture provides a rich representation that is sufficient for producing high quality image reconstructions in low-light and dynamic scenes with fast motion, the raw data clearly contains information that may improve object detection or classification pipelines. Data driven approaches that learn to solve inference tasks using single-photon data has been shown for both CIS-QIS[9] and SPAD arrays[10]. It is important to note that "photon-starved" regimes could include low-light, but also extremely small integration times. Low integration times could enable object detection in transient environments, increasing the effective speed of vision based manufacturing quality assurance or mobile robot platforms.

ACTIVE ILLUMINATION: LIDAR AND SEEING AROUND CORNERS

Active imaging incorporates the use of a controlled light source, such as lidar and flash photography. In recent years, SPAD arrays have shown significant traction in industry for use in ToF ranging, and are poised to continue to find adoption in industries that require compact and reliable 3D imaging.

Flash Lidar

A purely solid state lidar design using SPAD arrays known as Flash Lidar combines a pulsed laser source that illuminates a scene and timing electronics that timestamp photons as they arrive at the SPAD array after being reflected back from the environment. The SPAD array is placed behind a lens that images the environment, essentially creating a camera where each SPAD pixel corresponds to a particular direction into a scene. This SPAD camera has a field of view due to the imaging optics, and a pulsed laser source is used to illuminate the scene within the field of view of the SPAD camera. In order to maximize light efficiency, a VCSEL array is often used as the illumination source.

Seeing Around Corners

While lidar recovers the first bounce ToF of an active illumination source, the timing electronics often produce a histogram of returning photons. This histogram represents photon counts for buckets of time delays from the laser synchronization signal. While the peak count can be used for ranging, the full histogram includes a rich set of information, including multi-bounce information. When a laser pulse illuminates a scene, some light that is not scattered back into the SPAD camera from the line of sight surface can bounce around before returning to the SPAD camera. Effectively, when multi-bounce is present, the histogram can have multiple peaks, or appear smeared out over time. Conventional lidar tries to filter out this data, but non-line of sight imaging attempts to use the full histogram to produce ranging estimates beyond the line of sight. While a thorough review is not possible in this paper, significant attention has been paid by the computational imaging community to non-line-of-sight (NLOS) imaging in recent years[11].

As seen in Figure 1, real-time capture and reconstruction has recently been demonstrated in the lab. A picosecond pulsed laser is raster scanned on a relay wall, and a SPAD camera is used to capture histograms for each SPAD pixel focused on the same surface. The histograms are then used in a reconstruction algorithm to produce 3D image data from the perspective of the wall. While this enables imaging around corners, another way to understand this technique is that a virtual 3D camera can be projected onto a surface.

While flash lidar provides a compact 3D imaging solution, it is still limited to the line of sight. Using NLOS, 3D imaging can include parts of the world that are otherwise occluded by objects. For example, if a parked vehicle is occluding a pedestrian that is about to walk across the road, a lidar has no ability to detect the person. If a NLOS system was mounted on a moving vehicle, it could project a pattern on the road surface in front of the moving vehicle, enabling 3D imaging from the perspective of the projection. As the projected pattern moved within the line of sight of the pedestrian, the pedestrian would be detected well before a direct line of sight system would be able to see them.

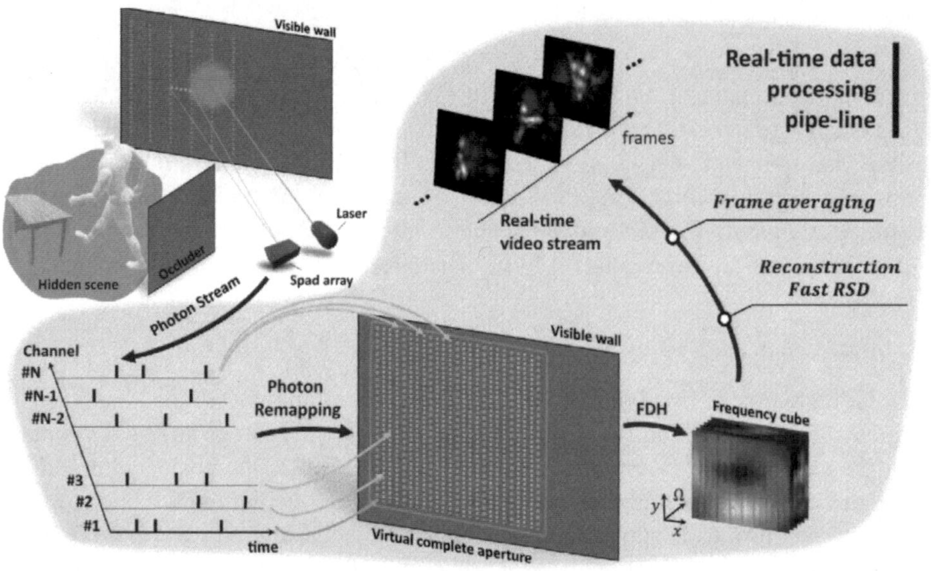

Figure 1. Non-line-of-sight imaging using time of flight has been demonstrated in the lab at 5 fps. A SPAD array observes a visible wall and generates ToF histograms with a synchronized pulsed laser. The ToF histograms are used to reconstruct an image of the scene from the perspective of the visible wall. [Figure from Nature Communications, Nam et al[12] under CCv4: http://creativecommons.org/licenses/by/4.0/]

DISCUSSION

Low Light and Fast Motion

When observing a scene with low light, CIS typically increase the exposure time in order to capture enough photons to form an image with a high enough signal to noise ratio (SNR). However, when there is motion, increased exposure times produce severe motion blur. In practical applications, motion blur can make computer vision tasks such as feature point tracking or face detection impossible. Photon-counting perception enables downstream processing to make use of a different type of input data. Rather than a sufficiently high SNR image with motion blur, single-photon perception makes use of photon-count cubes: a stack of frames each with photon counts and short exposure time. Due to the extremely low read-noise of single-photon data, these photon cubes can be integrated computationally, rather than selecting a long exposure time before image capture. Image stacks could be thought of as burst photography using CIS, but the read noise in these sensors accumulates when averaging many frames, creating an unacceptable decrease in SNR for downstream tasks when integrating more than a few frames at a time. The opportunity to integrate photon counts in diverse ways after image capture makes it possible to resolve fast motion in low light.

Dynamic Range

Imaging that makes use of photon counts promises to be inherently good at high dynamic range imaging. If photon counts are immediately digitized there is no fundamental limiting factor, such as full-well capacity for CIS, that leads to hard saturation of light flux estimates. In practice, single photon sensors can still saturate as frame-based methods still have a notion of an integration time where high flux levels will almost always lead to a photon detection. There is a similar problem for inter-photon timing methods available to SPADs, as the effective timing resolution is limited by clock resolution. Despite these limitations, photon counting remains a promising paradigm for improving the dynamic range of captured imagery.

Challenges for Single Photon Perception

There are a number of challenges for single-photon perception. Since much of the benefit of photon counts are due to the new computational possibilities, making use of single-photon data requires significant bandwidth and computation power compared to the analog exposure in CIS. As the cost of storage and compute continues to fall, new powerful applications made possible by single photon perception may justify the increased computational resource requirements.

A related challenge is power consumption. While single photon capture requires low energy for both QIS-CIS and SPAD, bandwidth and power requirements scale with the number of photons detected. SPAD arrays are particularly power hungry when avalanches are continuously triggered in high photon flux situations. For photon starved environments, this is not a problem, but for dynamic scenes with occasionally high photon flux, clever algorithms and sensor control methods must be employed to reduce power requirements. Despite these challenges, there are no fundamental roadblocks to continued adoption of single-photon image sensors for general perception applications.

CONCLUSION

Single-photon perception is at a turning point which has the potential to revolutionize optical perception. New sensing methods make it possible to use these systems to see around corners and in the dark, but must demonstrate that they can scale to become competitive with CIS in terms of image quality, cost, and compute. If the full potential of photon counting sensors and corresponding processing algorithms are realized, vision-based perception will increase in reliability and ability to operate in challenging environments while still being able to leverage decades of research in computer vision.

ubicept

info@ubicept.com
https://www.ubicept.com/

REFERENCES

[1] Sizhuo Ma, Shantanu Gupta, Arin C. Ulku, Claudio Bruschini, Edoardo Charbon, and Mohit Gupta. 2020. Quanta burst photography. *ACM Trans. Graph.* 39, 4, Article 79 (August 2020), 16 pages. DOI:https://doi.org/10.1145/3386569.3392470

[2] A. C. Ulku *et al.*, "A 512 × 512 SPAD Image Sensor With Integrated Gating for Widefield FLIM," in *IEEE Journal of Selected Topics in Quantum Electronics*, vol. 25, no. 1, pp. 1-12, Jan.-Feb. 2019, Art no. 6801212, doi: 10.1109/JSTQE.2018.2867439.

[3] T. Seets, A. Ingle, M. Laurenzis and A. Velten, "Motion Adaptive Deblurring with Single-Photon Cameras," in *2021 IEEE Winter Conference on Applications of Computer Vision (WACV)*, Waikoloa, HI, USA, 2021 pp. 1944-1953.
doi: 10.1109/WACV48630.2021.00199

[4] A. Ingle, *et al.*, "Passive Inter-Photon Imaging," in *2021 IEEE/CVF Conference on Computer Vision and Pattern Recognition (CVPR)*, Nashville, TN, USA, 2021 pp. 8581-8591.
doi: 10.1109/CVPR46437.2021.00848

[5] Fossum ER, Ma J, Masoodian S, Anzagira L, Zizza R. The Quanta Image Sensor: Every Photon Counts. *Sensors*. 2016; 16(8):1260. https://doi.org/10.3390/s16081260

[6] Chi, Y., Gnanasambandam, A., Koltun, V., Chan, S.H. (2020). Dynamic Low-Light Imaging with Quanta Image Sensors. In: Vedaldi, A., Bischof, H., Brox, T., Frahm, JM. (eds) Computer Vision – ECCV 2020. ECCV 2020. Lecture Notes in Computer Science(), vol 12366. Springer, Cham. https://doi.org/10.1007/978-3-030-58589-1_8

[7] A. Gnanasambandam and S. H. Chan, "HDR Imaging With Quanta Image Sensors: Theoretical Limits and Optimal Reconstruction," in *IEEE Transactions on Computational Imaging*, vol. 6, pp. 1571-1585, 2020, doi: 10.1109/TCI.2020.3041093.

[8] O. A. Elgendy, A. Gnanasambandam, S. H. Chan and J. Ma, "Low-Light Demosaicking and Denoising for Small Pixels Using Learned Frequency Selection," in *IEEE Transactions on Computational Imaging*, vol. 7, pp. 137-150, 2021, doi: 10.1109/TCI.2021.3052694.

[9] C. Li, X. Qu, A. Gnanasambandam, O. A. Elgendy, J. Ma and S. H. Chan, "Photon-Limited Object Detection using Non-local Feature Matching and Knowledge Distillation," *2021 IEEE/CVF International Conference on Computer Vision Workshops (ICCVW)*, 2021, pp. 3959-3970, doi: 10.1109/ICCVW54120.2021.00443.

[10] B. Goyal and M. Gupta, "Photon-Starved Scene Inference using Single Photon Cameras," *2021 IEEE/CVF International Conference on Computer Vision (ICCV)*, 2021, pp. 2492-2501, doi: 10.1109/ICCV48922.2021.00251.

[11] T. Maeda, G. Satat, T. Swedish, L. Sinha, R. Raskar, "Recent Advances in Imaging Around Corners, " ArXiv (2019).

[12] Nam, J.H., Brandt, E., Bauer, S. *et al.* Low-latency time-of-flight non-line-of-sight imaging at 5 frames per second. *Nat Commun* **12,** 6526 (2021). https://doi.org/10.1038/s41467-021-26721-x

Title: Automated Condition-Based Maintenance (CBM) Using Artificial Intelligence (AI) and Machine Learning (ML) for Unmanned Systems

AUVSI/Xponential 2022 - Intersection with Data XPO22-4076

Authors: Craig Proulx, eWorthy Technology Solutions, Inc and Karl Reichard, Ph.D. Penn State University/ARL

Abstract

As unmanned systems advance in capabilities, there are some aspects or components of this technology that lags behind and has a hard time catching up. It's been evident that the maintenance techniques and sustainment of unmanned systems has not kept pace with all of the autonomous technological advancements. As militaries, municipalities, and commercial unmanned assets multiply into larger and more complex fleets of assets there is now a demand for increased reliability and endurance. Going from one or two unmanned systems to a fleet of autonomous aerial, land and water assets, you create much larger operational risks and financial obligations for yourself. Once your mission develops into longer duration periods and distance, reliability and endurance needs increase in a no-touch contested environment. You have now created additional risk factors that are more complex and challenging than the unmanned system itself. The Department of Defense and commercial companies are now in a catch-up mode to increase reliability, endurance while decreasing sustainment life-cycle costs. Availability and endurance are needed while reducing life cycle sustainment costs. The Congressional Budget Office (CBO) indicated the U.S. Government budgeted over $6 billion dollars for AI in fiscal year 2022, but less than 4% will be used for manned and unmanned sustainment technology. Scheduled and planned maintenance policies continue to be the norm. Old maintenance practices can't and will not keep pace with increased availability and endurance requirements and will only increase sustainment costs. Current maintenance, logistics/supply systems and life-cycle engineering processes typically lack the intelligence to calculate remaining life expectancy and determine when parts will fail. They lack an integrated data environment that permits the prediction of optimal system performance, logistic support, and maintenance needs.

So, what's the answer and where is the CBM technology headed? Through current remote monitoring technologies, original equipment manufacturers (OEMs) are attaching sensors to many of the unmanned components and sub-systems, creating a clear path forward toward using CBM/Prognostics Health Management (PHM) techniques. These techniques have allowed for some unmanned systems to collect "Smart" data related to system health, performance, and environmental factors. This allows for advance automated analysis or data analytics to better diagnose unmanned systems and even help predict failures and provide preventative maintenance actions before the system actually fails. The objectives of CBM+ and PHM, are similar for both manned and unmanned platforms. In the traditional model of operations and sustainment, a fault would lead to a functional failure. In response, the system was taken out of service so repairs could be made before returning the system to service. In the case of an unmanned system, the system may have built-in redundant systems that would allow the remote operator to place the system in safe mode, diagnose the problem, and switch to a backup system before returning to service or aborting the mission. The addition of PHM allows early identification of faults and assessment of system capability and prognosis of remaining useful life. This provides a time horizon over which the operator or autonomous control system can modify or reconfigure the mission to extend the life, switch to backup systems, or reallocate the system to ensure success at the mission level.

Today, the same approaches to developing data-driven models for health assessment, CBM+, and PHM are also being used to develop digital twin representations of systems that can be used at the mission or fleet level. These digital twins can be used to compare current conditions and health to a known healthy baseline or updated with current information to assist in operational or logistics planning.

The need to collect, interpret, and act on data being broadcasted through remote monitoring from the sensors is critical to constructing a true automated CBM+ system. Unmanned systems using Artificial Intelligence (AI) and Machine Learning (ML) can receive, interpret, and process this sensor data in a data analytics construct to diagnose and predict failures, estimate spare parts and logistic resources. AL and ML can now increase endurance and tell you the life expectancy of your system allowing you to make critical real-time decisions before, during and after an operation or mission. Defense Departments and commercial companies need to approach a more advanced maintenance concept that uses AI and ML within a model-based systems engineering (MBSE) construct. AI can support CBM+ to meet maintenance free operational periods and reporting in an austere no-touch, contested environment. Without the correct CBM+ and PHM initiatives, owners and operators of unmanned fleets will not be able to accurately monitor and forecast operational availability. ML is needed for integrated diagnostic prediction and real time risk assessment. In order to frame this discussion, we need to separate AI and ML slightly. ML is a subset of a much larger framework for AI, so that ML refers specifically to algorithms and techniques that learn models from data and AI refers to techniques that reason and make decisions based on data, to include the output of ML models. Let's continue to investigate how AI and ML can be used to achieve increased endurance and help assess the life expectancy of your system allowing you to make critical real-time decisions before, during and after an operation or mission.

If we consider the flow and transformation of information in a CBM system, we can use models like ISO13374 and OSA-CBM, the Open System Architecture for Condition Based Maintenance. ISO 13374 defines six blocks of functionality in a condition monitoring system, as well as the inputs and outputs for each block. OSA-CBM is an implementation of the functional specification and defines data structures and interface methods between blocks. The functional blocks include:

- Advisory Generation (AG)
- Prognostics Assessment (PA)
- Health Assessment (HA)
- State Detection (SD)
- Data Manipulation (DM)
- Data Acquisition (DA)

These functional blocks also map to levels of data refinement and processing. When ISO 13374 and OSA-CBM were developed in the 2000s, AI and ML techniques were not as well known or widely used as today, and the processing models were primarily physics based or relied on comparing signals and derived quantities to thresholds based on the experience of subject matter experts.

Let's take a look at where AI and ML can play a role in the implementation of this model for machinery condition monitoring and control or action. The lowest levels of the hierarchy, DA and DM, rely primarily on traditional signal processing techniques: filtering data, calculating statistical measures, spectral analysis, and other data transformation. State Detection refers to the calculation of condition indicators associated with specific fault conditions and the detection of conditions associated with outside known normal or health operating conditions or anomalous behavior. State Detection and Health Assessment rely on classifying data as representing a healthy condition, or one of many potential faulty conditions (e.g. outer race fault on the drive motor output shaft bearing). While AI and ML are hot topics now, we have been using machine learning for many years or decades to build pattern recognition and classification systems across many problem areas, including CBM. The difference today is in our ability to combine techniques and computations to handle many more inputs to the models. Machine learning is routinely used to build models to detect faults and classify machinery health, including techniques such as classic neural networks, nearest neighbor classification, principal component analysis, and other variations on these themes. Prognostic assessment relies on building regression models that can predict the evolution of data and condition indicators to provide insight into when the condition will degrade to the point where it will impact mission capability.

The ability to determine the consequence of faults and failures and the course of action required to mitigate the effect of the degraded condition, how to proceed with the mission, and how to repair or return the asset to full mission capability is where we still rely primarily on trained operators, planners, and maintainers. While we are still way off from having robots autonomously repairing the equipment, it is not unrealistic to consider the use of AI to assist in responding to degraded conditions, replanning or redirecting the mission, and ordering or scheduling repairs and parts to return the system to full mission capability.

Now let's look at some of the opportunities and challenges in applying AI and ML for condition monitoring in autonomous vehicles. First, let's look at opportunities. As mentioned before, we have been using machine learning techniques to build signal detection and classification system well before the latest upswing in the AI/ML hype cycle. What's different now? The first thing is the increase in computational capability and efficient implementation of neural networks. 10-15 years ago, we were limited to 3-layer neural networks and tried to minimize the number of input features used in fault classification. This applied not only to neural networks, but also most other approaches. Now, however, we routinely see neural network classifiers or regression models with 10s, 100s, or 1000s of inputs (think how many pixels are in a typical photograph) and train those models on millions of data points. Increased computational capability (and memory) is just one factor, however, since it still relies on human subject matter experts to decide what to use as inputs to the models (the output of the data manipulation and state detection layers described previously). In this case, we are still limited by not knowing what we don't know about system interactions that aren't described by the physical and engineering models we SMEs are trained on. The advantage that the improvements in computation capability does provide is that we can include many more variables in the model. For example, if I want to detect and predict the evolution of that bearing fault mentioned earlier, can include traditional physics and engineering-based condition indicators based on vibration measurements, but you can also include the time of day, weather conditions, operator information, number of operating hours, and many other data that might not have considered relevant given the physics. One of the promises of ML is learning dependencies that may not be obvious.

The other big revolution in ML is the development of deep learning techniques. We studied publications in the area of deep learning for turbine engine CBM several years ago and you can trace the steady growth in the area over the last 15 years. One of the most recognized applications of deep learning is in image recognition. The development of deep learning on CBM is twofold. First, it allows us to cast detection and classification problems as image recognition problems. For examples, human SMEs examine plots of signals versus time, or time-frequency plots (signal energy as a function of time and frequency) to determine patterns and make detections and classifications. We can take those same "images" that human SMEs use and train a deep learning neural network to do the same, in the same way we can train them to tell cats from dogs, or the way your phone or computer can automatically label friends and family members in photos. Buried in the details of how the deep learning neural network works, however, is another feature of the approach that is important for the application to CBM. During training, the network not only learns the difference between cats and dogs, but it also learns the features that distinguish the two, without the subject matter expert trying to describe the difference ahead of time and doing what is known as feature engineering. The potential advantage here is that the network encodes knowledge from the data internally that may detect hidden dependencies or indicators that we cannot conceive. The ability to rapidly train and implement high dimensional neural network models is so important, that chip makers are now designing hardware optimized for the execution of the training and processing algorithms. This goes beyond just using multiple processing cores of repurposing the processing architecture of GPUs.

While AI and ML hold much promise, we should also be sober about the challenges and limitations. The first is the requirements to train the neural networks. Neural networks require a lot of data to train, and big neural networks, with large numbers of internal nodes and connections need commensurately more data. Self-driving cars have autonomy algorithms trained to recognize stop signs, crosswalks, bicyclists, and other objects. How do they do that? They use labeled data in a construct called supervised learning, meaning the neural network is fed a diet of images that are labeled with the objects they contain. How do they get those labels? Have you ever had to verify that you weren't a computer by selecting all of the images in a grid with fire hydrants? You were helping to train these models. We don't have that in the maintenance world. We have several problems. First, we have an unbalanced set of data from healthy and degraded systems, with (fortunately) the preponderance of the data corresponding to health systems. Second, most of the data is not reliably labeled. The good news, however, is that the same deep learning networks that was described before can build an internal representation of a good or healthy system and compare incoming measured data to the good data model to detect anomalous conditions. There are also new techniques for using unsupervised learning to cluster data and then assign whole clusters to classes instead of individual data records, reducing the burden of labeling data.

A second big challenge is the prediction of asset health and the impact on the mission. Predicting future health is a function of current health or system state, the physics of how faults progress within the system, and the future demands and loads. In most, but not all, cases, if you don't use the system, the faults won't progress, and the system health will not change. However, the health at some time in the future (say 1 week) may depend significantly on how you use the assets or the type of mission you execute over the next several days. The point is that in order to accurately predict future health, the health monitoring system must be integrated with many other onboard and off-board systems. Traditional condition monitoring systems do not have that level of integration and rely on a human intermediary to understand the impact of current and planned operations on the progression of faults or degraded conditions. This is not only an integration issue, but a challenge for the larger AI realm in building intelligence into the system that can perform this type of reasoning.

The final challenge worth noting is one of system integration. This goes beyond the integration of information required to make predictions that was just described. Here we are talking about the language of machinery condition, controls, logistics, and maintenance. Today, we have health monitoring systems on many platforms, but they are designed to "talk" to human operators – engine oil temperature is high, "check engine." We also have autonomous control systems capable of interpreting sensor signals and navigating between waypoints or around obstacles. In order to have the autonomous controller operating the system also act on health information, the autonomous control system and the health monitoring system must speak a common language and have a shared internal model of the platform. Because these two systems have traditionally been developed in parallel or at various times in the system design lifecycle, they often don't talk the same language and still rely on a person to translate and recode the information. There are efforts underway to build system models that account for the different roles in system operation. An example would be a human-centric crew model where the autonomy system is comprised of agents acting as the commander, the engineer, the logistician, etc. An example of this is the Unmanned Maritime Autonomy Architecture.

Conclusions: We can think about the use of CBM to inform unmanned autonomous operation by imagining a Venn diagram. One circle or shape described the capability of an asset or system based on the original design, any sensors or other payloads, and an assumption of operating in a perfectly healthy state. As the system ages, or specific components degrade or fail, this capability space shrinks to something inside the original capability space. When we plan a mission, we are defining a requirement in this capability space that we expect the system to be able to meet, which defines another shape. If the mission capability is contained entirely within the current capability space, then we have a high probability of mission success. If, however, the mission capability requirements extend beyond the boundary of the current capability space, then the probability of mission success decreases because we cannot meet those requirements. The role of AI and ML in this context is to identify the current health, the effect of the planned mission on the system capability and adjust the mission execution to ensure the required capability remains within the boundaries of what the system can achieve. Post mission, we want the AI system to have already requested resupply, parts, and maintenance to quickly restore full system capability.

RIGID FORMATION CONTROL APPLIED TO UNMANNED SURFACE VEHICLES

Jennifer T. Nguyen,[1] Matthew Feemster,[2] and Paul Frontera[3]

A graph rigidity formation control (RFC) is investigated for use in maneuvering a formation of marine vessels along a desired bearing and non-zero speed while simultaneously maintaining a commanded inter-vehicle spacing. In an effort to apply to a wide variety of vessels and also to reduce stress on the vessel's drive train, the motion of the vessel is restricted to be in its corresponding ``forward'' direction. An inter-vehicle error allowance is employed in order to accommodate the forward velocity constraint. Simulation results of three scaled vessels operating in formation are provided.

INTRODUCTION

As sensor and processing capabilities continue to grow, the ability to deploy autonomous unmanned vehicles (AUVs) also increases. Networks have developed to become stronger and much more secure, allowing for AUVs to travel farther. The current applications for uncrewed surface vehicles (USVs) include navigational aids, climate monitoring and data collectors, surveyors, tugs, and military operation tools.[1] They are more useful and versatile than weather buoys, but cheaper and more agile than weather research ships.

Unmanned surface vehicles provide a wide array of capabilities and services. One of the major applications is for hydrographic surveying. Hydrographic surveys collect data from multi-beam sonar readings which measure the depth of the sea floor and other waterways using sound waves paired with global positioning system (GPS) data.[2] The National Oceanic and Atmospheric Administration (NOAA) relies on hydrographic surveys to create and update nautical charts. These navigational charts serve commercial, military, and recreational mariners alike. They ensure that shallow features, shorelines, and other navigational hazards are known. Hydrographic surveys are also conducted to determine water quality and marine life population. The mobility and flexibility of USVs also allows for deployment in sonar technology to conduct search and rescue operations.

To enable USVs to carry out these missions, there needs to be a program capable of commanding each vehicle's speed and bearing. An algorithm to maintain a fixed-distance formation has been created and implemented onto unmanned ground vehicles. The same algorithm would be unable to be directly applied to USVs. The vehicle dynamics and mode of mobility completely differ. Unmanned surface and fixed wing aerial vehicles are unable to drive backwards in the same way ground vehicles do. Surface vessels in water also have much larger turning radii. The momentum from the thrusters will cause the vehicle to continue to turn even after the motors are shut off. Resilience against environmental factors is another requirement that need to be considered and

---

[1] Midshipman, Robotics & Control Eng. Dept, U.S. Naval Academy, jennifertaylornguyen@gmail.com.
[2] Professor, Robotics & Control Eng. Dept, U.S. Naval Academy, feemster@usna.edu.
[3] Permanent Military Professor, Robotics & Control Eng. Dept, U.S. Naval Academy, frontera@usna.edu.

tested (i.e. current or wind). In the future, the refined algorithm discussed in the paper can then be applied to other uncrewed vehicles with restricted maneuverability.

BACKGROUND

The primary objective of this research project is to investigate the feasibility of applying the graph rigidity formation control (RFC) of approach[3] to a swarm of marine vessels. Graph theory is a convenient tool for describing the multi-agent formation shape as well as the inter-agent sensing, communication, and control topology in a decentralized scheme when the main concern is the convergence to a desired spatial configuration irrespective of its exact, global position. However, since the approach is based on the assumption that each agent is a holonomic vessel, fielded implementation may not yield the expected performance due to improper compensation of the dynamic effects influencing each agent's motion (*e.g.*, inertial, frictional, gravitational, non-holonomic, actuator, *etc.*)[3]. The RFC was initially experimentally evaluated in Reference 4 on three scaled *electric* ground vehicles shown in Figure 1.

Figure 1. Three scaled electric vehicles for RFC evaluation.

From Figure 2, two primary observations are gleaned: i) as the formation approaches the desired inter-vehicle spacing value, the RFC algorithm commands ``high" frequency velocity commands to maintain the desired distance between the vehicles and ii) the electric vehicles are able to produce symmetric forward/reverse velocity motions.

For application to a group of surface vessels, the above issues will prove to be problematic. Similar to the study in Reference 3, three scaled autonomous surface vessels will be employed to evaluate the rigid formation control algorithm's implementation for a maritime application.

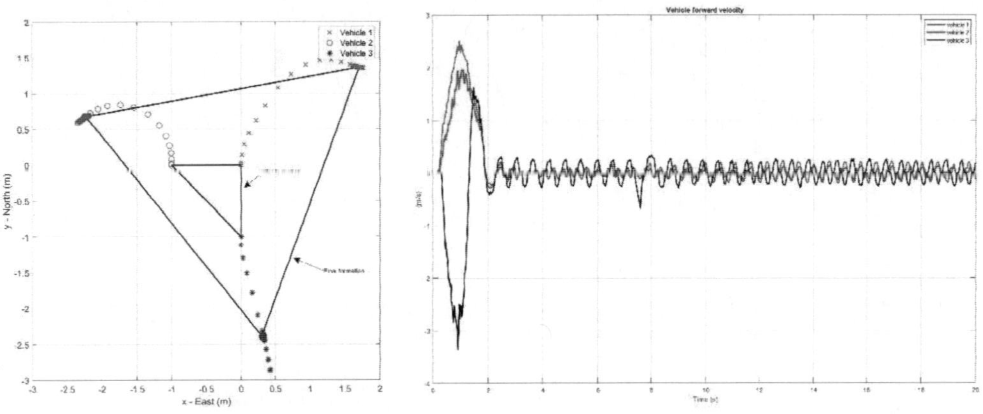

Figure 2. (a) Formation acquisition (b) vehicle velocities.

The Clearpath Robotics Kingfisher vessels utilize an inboard propeller to generate thrust. For positive thrust, water is pulled up through an intake on the bottom of the pontoon and then propelled directly out the rear. As a result, the Kingfisher's thrust generation profile is shown in Figure 3.

Figure 3. (a) Formation acquisition (b) vehicle velocities.

Due to the imbalance between the generated forward/reverse thrusts observed in Figure 4, the Kingfisher surface vessels will experience difficulty (as compared to the electric ground vehicles) in reproducing the balanced positive/negative velocity motion when the vessels approach the commanded inter-vehicle distance. In an effort to avoid transitioning between forward/reverse thrust and to appeal to a broader class of vehicles such as gas-powered outboard vessels or other common marine vessel powertrains, the forward motion of the surface vessels will be restricted to only positive velocities. Specifically, the surface vessels will adjust their forward speed and heading to promote the inter-vehicle distance objective and the formation's objective (*i.e.*, a desired formation bearing and desired formation velocity) and not command astern propulsion.

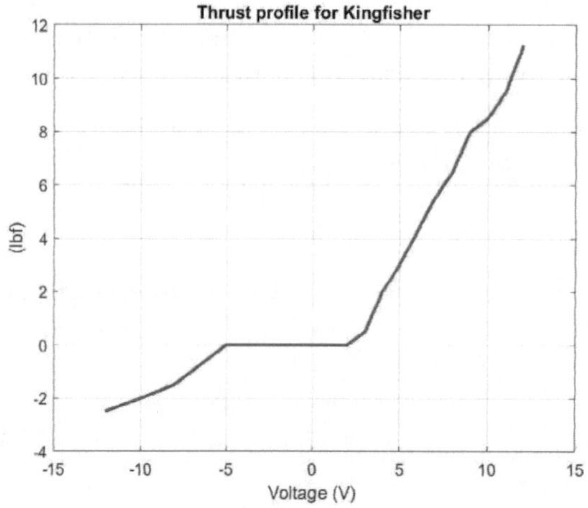

Figure 4. Kingfisher Thrust vs. Applied Voltage.

SIMULATION REVIEW

In the subsequent simulations, three Clearpath Robotics Kingfisher vessels (see Figure 3) are commanded to develop an equilateral triangle formation with an inter-vehicle distance of 7.0m. In addition, the three-vehicle formation is to transition along a bearing of 123.0° at a rate of 0.75 m/s as summarized in Table 1. The proposed RFC algorithm only utilizes the relative position to each formation member (*i.e.*, not velocity or heading).

Table 1. Formation Specifications.

Desired formation bearing	123.0°
Desired formation velocity	0.75 m/s
Desired inter-vehicle distance	7.0 m

The RFC algorithm commands a North-East-Down (NED) velocity command for each vessel based on the inter-vehicle distance and the overall formation's specifications of Table 1. For each simulation, the commanded NED velocity signals are then directly converted into a reference heading and reference forward velocity command for each vessel to achieve.

Simulation: RFC Unmodified

For this simulation, the non-linear thrust characteristics of Figure 4 were integrated and each vehicle's velocity is constrained from requiring astern propulsion. Figure 5 illustrates the formation's path. The observed loops and wide turns are a result of the imposed forward motion constraint and the limited negative thrust. From figure 5, it is obvious that direct application of the RFC with the forward velocity constraints fails to offer a viable formation.

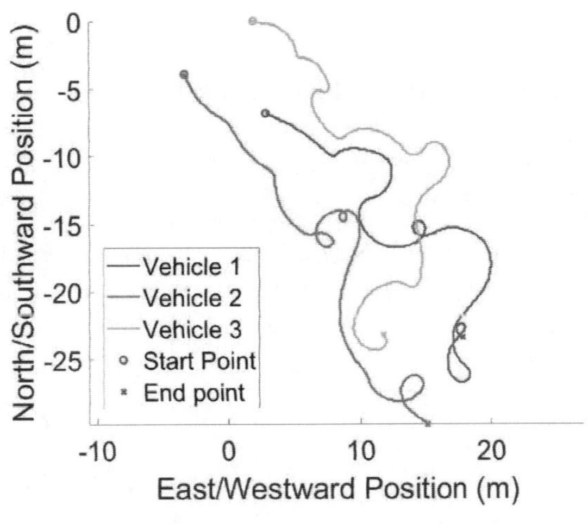

Figure 5. Path of all simulated Kingfishers with distance error allowance.

Figure 6 displays the vehicle inter-spacing distance. The oscillatory nature of the inter-spacing distance can be attributed to each vehicle's restriction to turn “about" as opposed to reversing as well as the rigid formation controller's proportional foundation (*i.e.,* the RFC utilizes a proportional control scheme to achieve the inter-vehicle range objective).

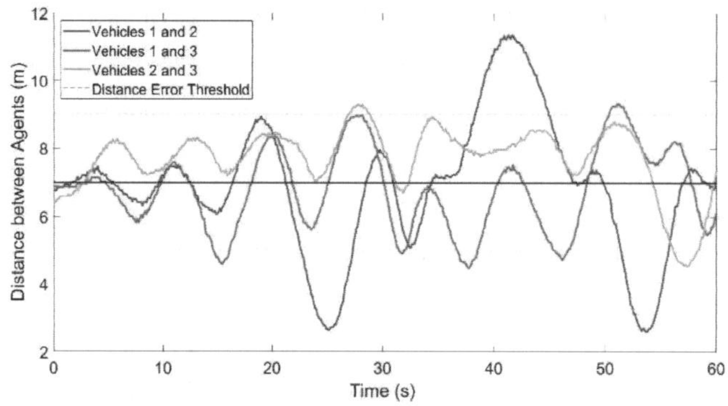

Figure 6. Distance (m) between of all simulated Kingfishers with no distance error allowance.

Figure 7 demonstrates that each member of the formation's velocity is always positive. Vehicle 1 for example, dips down to zero to allow the other agents to catch up but it doesn’t travel in a negative direction.

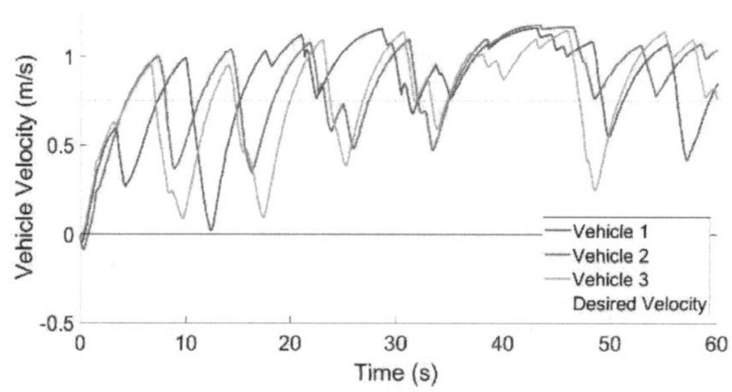

Figure 7. Velocity (m/s) of all simulated Kingfishers with no distance error allowance.

Simulation: RFC with Error Allowance

In order to improve the acquisition of formation with the marine vessels, an error allowance is incorporated into the rigid formation control algorithm. Specifically, if the inter-vehicle spacing error between two agents is within a specified allowance then the RFC's overall commanded velocity is solely based on the formation's desired speed/heading objective. In this simulation, a ± 1.5m error allowance value is utilized. Figures 8 and 9 illustrate the resulting formation path and inter-vehicle distance, respectively.

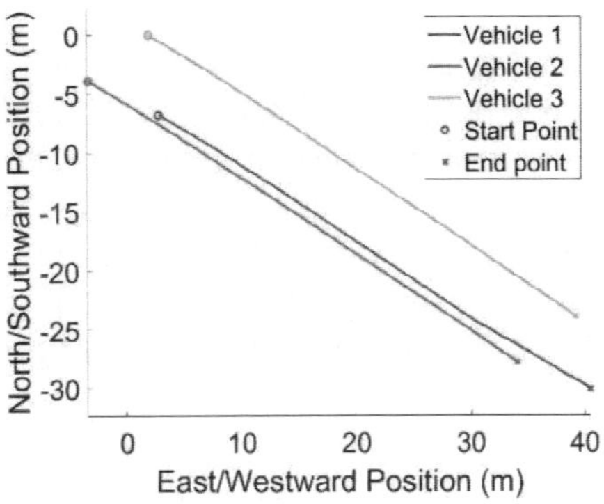

Figure 8. Path of all simulated Kingfishers with no distance error allowance.

From Figure 8, the erratic behavior of the vessels is diminished with the employment of the error allowance. In essence, the allowance offers the vehicles the ability to align the formation's desired heading/speed when within the error bound and then make minor heading/speed adjustments in order to stay within the error bound as seen in Figure 10.

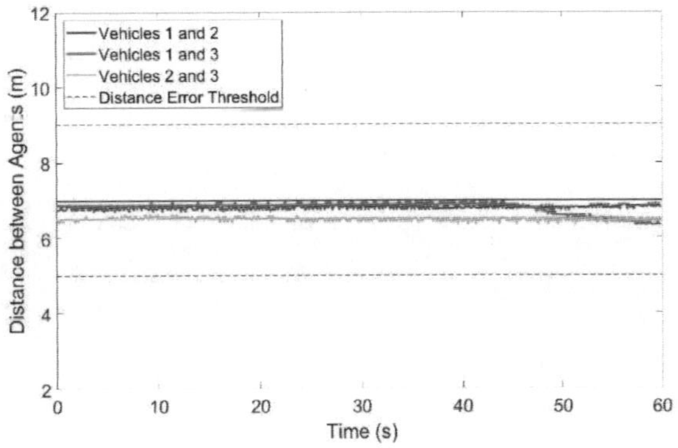

Figure 9. Distance (m) between of all simulated Kingfishers with distance error allowance.

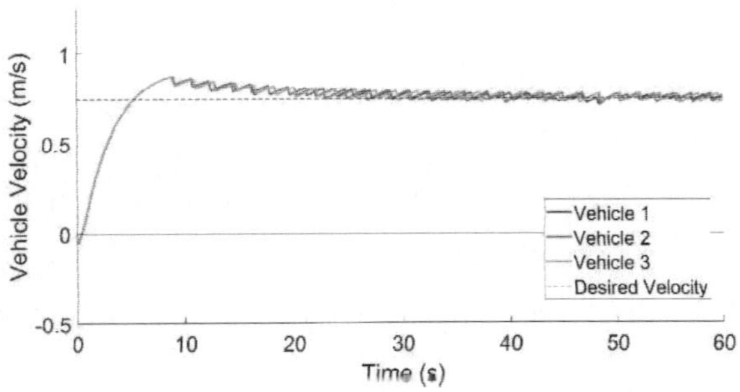

Figure 10. Velocity (m/s) of all simulated Kingfishers with distance error allowance.

CONCLUSION

In this paper, a graph RFC approach has been augmented to employ an error allowance/tolerance on the inter-vehicle distance signal to promote formation acquisition/transition of a group of marine vessels moving along a desired non-zero speed/bearing. The injection of the error tolerance allows for the influence of the vessel dynamics to be somewhat diminished thereby allowing the high-level kinematic RFC algorithm to promote formation acquisition while simultaneously allowing the members of the marine formation to maneuver only in the forward direction. The error allowance injection has been evaluated within a nonlinear simulation setting. Future work will be focused on the execution of fielded experiments.

REFERENCES

[1] "USV (Unmanned Surface Vehicle), applications and advantages," Embention, 18 Sep 2015. [Online]. Available: https://www.embention.com/news/usv-unmanned-surface-vehicle-applications-and-advantages/. [Accessed 4 Dec 2021].

[2] National Oceanic and Atmospheric Administration. "Hydrographic Surveying," NOAA's National Ocean Service. [Online]. Available: https://oceanservice.noaa.gov/navigation/hydro/. [Accessed 14 April 2022].

[3] De Queiroz, Marcio, Cai, X., and Feemster, Matthew. Formation of Control of Multi-Agent Systems. A Graph Rigidity Approach. 1st ed. Wiley, 2019.

ADVANCES IN ALD/PARYLENE TECHNOLOGY FOR ROBUST PROTECTION OF SENSORS AND ELECTRONICS

Rakesh Kumar* and Shuichi Sawada†

This research work describes an advanced ALD conformal coating, applied at room temperature, making it suitable for sensors, printed circuit boards, and other components used in unmanned vehicle, aerospace and military assemblies to provide robust and reliable protection from corrosion and other harsh environments. In addition to these beneficial properties, experiments demonstrate excellent corrosion resistance, adhesion to the substrate, electrical insulation and a significant (sixty-three times) improvement over existing technology of Parylene C's water vapor transmission rate when combined with ALD. The use of this technology may also help towards the goal of decreasing overall system weight compared to existing options for the unmanned vehicle, aerospace, avionics and defense industries.

INTRODUCTION

As technologies continue to evolve, the requirements for high level of protection and reliability of the assemblies, components and electronics used in aerospace and military assemblies, particularly in harsh operating conditions, are also increasing. In addition, light-weighting continues to be a key topic for the unmanned vehicle, aerospace, avionics and defense industries as new metals and composites are being integrated into end products and assemblies with the goal of decreasing overall system weight. While several liquid conformal coatings have historically offered a certain level of protection to components in these industries, these coatings add unnecessary weight and dimension to critical, lightweight electronic, sensors, optical and communication devices.

The Atomic Layer Deposition (or ALD) technique, which was introduced in late 1990's, has been advancing in many applications as it deposits ultra-thin films (a few nanometers) of metal and metal oxides with a precise thickness and conformality. Although initially started in semiconductor industry, ALD has recently gained a lot of interest as a nano-sized films or nano-sized coatings in various industries[1, 2, 3, 4, 5]. Based on sequential, self-limiting reactions, the thin films/coatings produced by ALD are highly conformal on high aspect ratio structures, and can be utilized on a wide range of geometrically complex surfaces, and even spherical particles. However, its use in general aerospace and military electronics has been limited due to its high temperature processing conditions.

One of the key factors is the varying environments around the world. Electronics are getting more affected by moisture, water and other corrosive chemicals. The corrosion of PCB assembly and their specific components is primarily due to salt water. As you may know, moisture finds its

* Vice President of Technology, Specialty Coating Systems, Inc. Indianapolis, USA

† Mechanical Engineer, Daisan Kasei Co., Ltd, Japan

way into the smallest areas and spaces of any electronic device leaving corrosive deposits behind that continue to cause damage after drying. It is, however, not simply "moisture" that does the damage; it is the ions such as Calcium, Magnesium, Sodium, Potassium, Chloride, Nitrates, Iron, Copper etc. in solution. Moisture in the air contains similar substances as well as particulates from air pollution, industrial contaminants, and chlorides in marine environments.

Currently there are many ALD methods available and each one has its own unique processing conditions and application potentials. The thermal ALD was the most common method initially, also called classical ALD, but now there are several processing methods based on low temperature and ease of energy source requirements. The temperature range utilized in thermal ALD is generally between 150-350 °C. For many substrates, high temperature is not suitable, hence other method such as plasma enhanced ALD or electron enhanced ALD are preferred. There are many details in each method, which is beyond the scope of this work. So, why ALD-Parylene for robust protection of electronics? Let us understand ALD and Parylene separately first. Parylene is the name for a unique series of polymeric organic coating materials that are polycrystalline and linear in nature, possess excellent dielectric and barrier properties per unit thickness, and are chemically inert. Parylene conforms to all surfaces, edges and crevices of a substrate, including the interiors of multi-layer electronic packages. Parylenes are RoHS, REACH compliant, and have been proven to provide metallic whisker mitigation in lead-free solder applications. Parylenes are ideal for protecting electrical components, wires, components of electric power drivetrains, PCBs, sensors and components that need reliable, long-life performance in harsh environments[6].

For ALD coating, an inorganic material, aluminum oxide was used, which is also a biocompatible coating and deposited at room temperature using plasma enhanced ALD process. Aluminum oxide ALD coating is highly conformal, atomically flat and smooth surface coating and it can be easily combined with organic coating. It has excellent barrier properties at Angstrom level, particularly against water vapor, ions and other gases. So now, one can understand that both aluminum oxide ALD coating and Parylene coating have some unique features to provide excellent barrier protection, but in each case, one has to apply individual coating at certain thickness to achieve the desired performance. However, when a robust and long-term protection is required, and higher thickness is not possible due to miniaturization of electronics operating at high frequency ranges, a combination of aluminum oxide ALD, and Parylene becomes the most suitable coating option at ultra-thin level to protect the electronics from corrosion and other environmental degradation[4, 5].

This research work demonstrates an advanced ALD conformal coating (in combination with Parylene technology) that is applied at room temperature, making it suitable for printed circuit boards and other components used in aerospace and military assemblies to provide robust and reliable protection from corrosion and other harsh environments. Additional data demonstrate excellent corrosion resistance, adhesion to the substrate, electrical insulation and a significant (sixty-three times) improvement over existing technology of Parylene C's water vapor transmission rate when combined with ALD. The use of this technology would also help towards the goal of decreasing overall system weight compared to existing options for the unmanned vehicle, aerospace, avionics and defense industries.

EXPERIMENTAL

Our experiment adopted two combinations, one with Parylene on top and other with Aluminum oxide ALD coating on top with excellent adhesion in both cases before subjecting to various tests as shown in Figure 1.

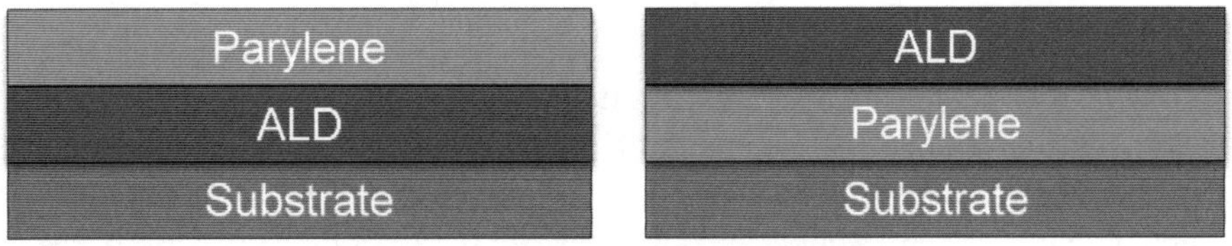

Figure 1. ALD Parylene Combination

Moisture Vapor Transmission Rate

Our first experiment was to determine the Water Vapor Transmission rate (WVTR), which relates to assessing the barrier performance of any coating. A polyethylene napthanate was used as a base material, with known water vapor transmission rate. diX-C, which is also called Parylene C, was applied after aluminum oxide ALD coating as shown in Figure 2. The measurement standard used was JIS K 7129 to determine the water vapor transmission rate.

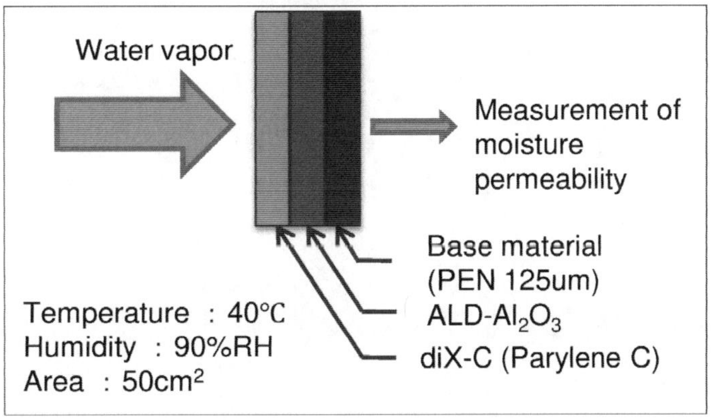

Figure 2. Water Vapor Transmission Rate (WVTR) Test Set Up

Gas Permeability

Following a similar hybrid system as for WVTR, we deposited 10-nanometer aluminum oxide ALD coating on the substrate, then applied 10 microns thick diX-C on top of the ALD coating, and tested gas permeability of oxygen and hydrogen. The JIS K7126 standard was used for gas permeability, which utilizes differential pressure method.

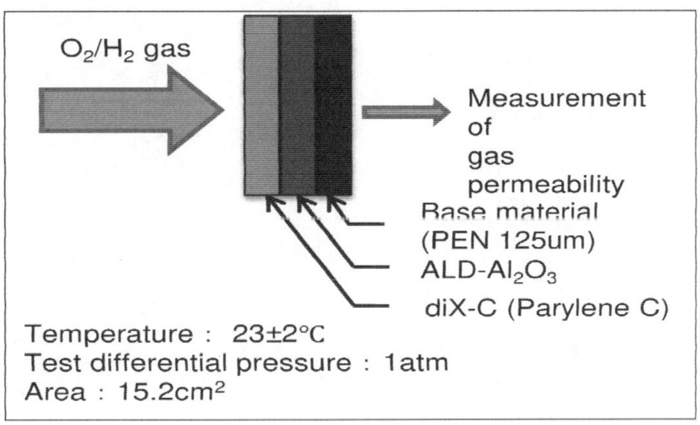

Figure 3. Gas Permeability Test Set Up

Corrosion Resistance

After the initial barrier and gas permeability testing, coated several steel plates with 10 nanometer thick aluminum oxide ALD coating as a first layer and with varying thickness, ranging from 0.5 to 5 microns of diX-C or Parylene C coating on top of the ALD coating. Also coated several samples with the varying thickness, ranging from 0.5 to 5 microns with diX-C coating only. The salt spray test included three samples of each thickness combination. Test continued for more than five hundred hours.

Figure 4. Corrosion Test Set Up

Surface Roughness

In order to understand the surface roughness and pinhole free nature of ALD coating, Atomic Force Microscopy (AFM) was used with nanoscale resolution.

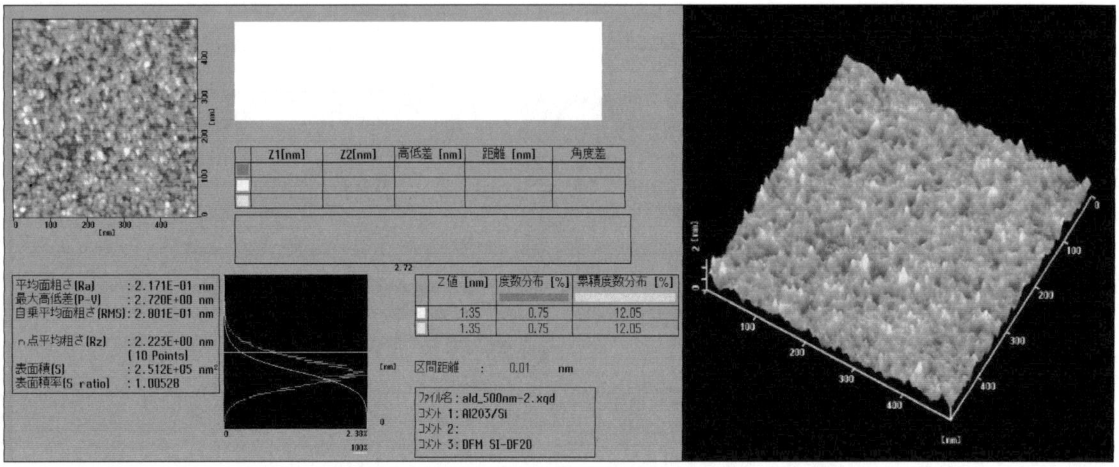

Figure 5. AFM Image Showing Surface Flatness

RESULTS AND DISCUSSION

As shown in Table 1, the diX-C or Parylene C alone has water vapor transmission rate of 0.19 g.mm/m^2.24 hours at thickness of 0.5 microns. When diX-C or Parylene C was combined with 18 nm thick aluminum oxide ALD coating, the water vapor transmission of this hybrid system became 0.003 g.mm/m^2 .24 hours, which is 63 times better. The goal of this study was to find the lowest possible thickness that can provide lowest (best) water vapor transmission rate. A low water vapor transmission rate indicates better barrier performance.

Table 1. Water Vapor Transmission Data

Deposition Condition	WVTR [g · mm/(m2 · 24h)] 40°C/90%RH
Reference (PEN film)	0.21
diX-C 25um	0.17
diX-C 10um	0.19
diX-C 5um	0.18
diX-C 0.5um	0.19
diX-N 10um	0.22
1st layer: Al_2O_3 10nm + 2nd layer:diX-C 10um	0.009
1st layer: Al_2O_3 10nm + 2nd layer:diX-C 0.5um	0.004
1st layer: diX-C 10um + 2nd layer Al_2O_3 10nm	0.026
1st layer: Al_2O_3 18nm + 2nd layer:diX-C 0.5um	0.003
1st layer: Al_2O_3 10nm + 2nd layer:diX-N 10um	0.010

The results of gas permeability testing are shown in Table 2. As indicated by the results, the combination of both the Parylene and ALD coating provided the best barrier performance against oxygen and hydrogen. For both oxygen and hydrogen, the permeability data is lower than detection limit of the measuring equipment. The composite property of diX-C and ALD-Al_2O_3 showed remarkable improvement in gas permeability property.

Table 2. Gas Permeability Data

Deposition Condition	Gas Transmission			
	(cm^3/m^2 · 24h · atm)		(mol/m^2 · s · Pa)	
	O_2	H_2	O_2	H_2
diX-C film (10um)	3.84	198	1.96E-12	1.01E-14
ALD-Al_2O_3 (10nm)	0.45	27.8	1.42E-16	2.76E-18
1st layer ALD-Al_2O_3 (10nm) + 2nd layer diX-C (10um)	< 0.05*	< 0.05*	< 5.00E-16*	< 5.00E-16*

*Regarding gas transmission, the oxygen/hydrogen gas data is lower than the detection limit of the equipment.

**Base material: PEN 125um

The results of Atomic Force Microscopy (AFM) indicate that the coating is flat with RMS value of 0.28 nanometer. Looking at the value Ra of 0.22nm, which indicates that the ALD coating is very dense. The AFM images show no sign of any cracks or pinholes.

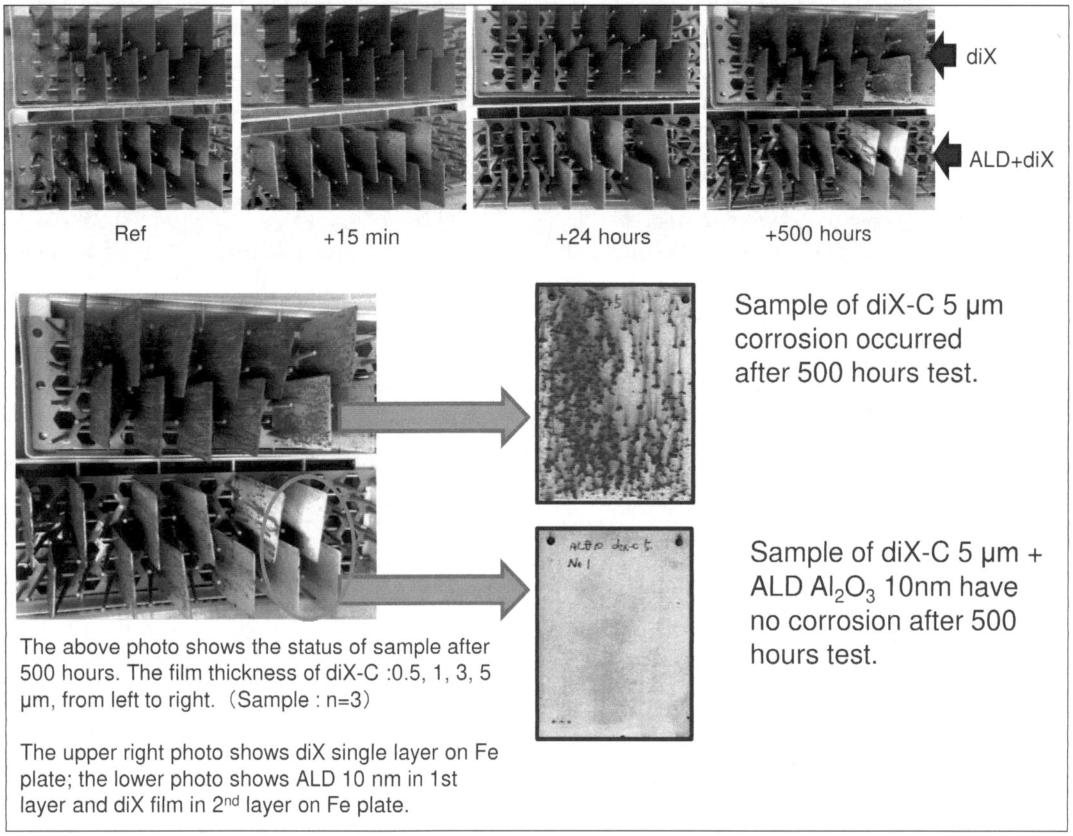

Figure 6. Corrosion Resistance Results

All samples coated with ALD and diX-C survived more than 24 hours of continuous spray without any corrosion while samples coated with lower than 5 microns of diX-C only showed corrosion after 24 hours. When examined after 500 hours of spray, all samples coated with diX-C only and samples coated with hybrid of ALD coating and diX-C coating where diX-C thickness was less than 5 microns showed corrosion. However, the hybrid samples that had 10 nanometer of ALD coating and 5 microns of diX-C or Parylene C coating had no corrosion after more than 500 hours. As shown in Figure 6 (the close up of the samples), the samples with 10 nanometer of ALD coating with 5 microns of diX-C coating showed the best resistance against corrosion in salt water harsh environment. Based on this harsh environment evaluation, it is clear that composite or hybrid coating of ALD with only 10 nm thickness and diX-C with at least 5 microns can provide best long-term protection for electronics against corrosion.

CONCLUSION

ALD coating, particularly aluminum oxide, processed at room temperature and combined with Parylene coating at ultra-thin level offers an excellent option for providing robust and reliable protection against various corrosive agents and environments. In addition, based on the barrier performance data, Parylene coating on top of the ALD coating provides best electrical insulation, protection against water splash and water immersion at ultra-thin level.

REFERENCES

[1] A. K. Singh, K. Adstedt, B. Brown, P. M. Singh, and S. Graham, "Development of ALD Coatings for Harsh Environment Applications", ACS Applied Materials & Interfaces, 11 (7), 7498-7509, 2019.

[2] R. Caldwell , H. Mandal, R. Sharma, F. Solzbacher, P. Tathireddy, L. Rieth, "Analysis of Al_2O_3-parylene C bilayer coatings and impact of microelectrode topography on long term stability of implantable neural arrays", J Neural Eng. Aug;14(4):046011. DOI: 10.1088/1741-2552/aa69d3. PMID: 28351998, 2017.

[3] A. Hogg, "Development and characterization of ultrathin layer packaging for implantable medical devices", Ph.D. Thesis, August 2014, Graduate School for Cellular and Biomedical Sciences, University of Bern, ISBN 978-2-940387-14-4, 2014.

[4] X. Xie, L. Rieth, S. Merugu, P.Tathireddy, and F. Solzbacher, "Plasma-assisted atomic layer deposition of Al2O3 and Parylene C bi-layer encapsulation for chronic implantable electronics", Applied Physics Letters 101, 093702, 2012.

[5] N. Kim, W. J. Potscavage Jr., B. Domercq, B. Kippelen, and S. Graham, A hybrid "Encapsulation method for organic electronics", Applied Physics Letters, 94, 163308, 2009.

[6] R. Kumar, "Xylylene Polymers", Kirk-Othmer Encyclopedia of Chemical Technology, John Wiley, New York, Online publication, 2008.

SENSITIVITY ANALYSIS OF NASA AFRC'S EXPANDABLE VARIABLE-AUTONOMY ARCHITECTURE

Noel Khan [1] and **Brandon Persons**[1]

NASA's Expandable Variable-Autonomy Architecture (EVAA) is an all-domain government-owned reference implementation of a type of supervisorial-control based on the ASTM 3267-17 standard for bounding the behavior of complex systems. EVAA was designed to address both regulatory and technical challenges in autonomy. As an extensible framework, EVAA plugins can be used for guidance, recovery, sensors, trajectory predictions, and vehicle control. This paper evaluates EVAA's roughly 100 parameters with respect to process contention, data distribution, trajectory predictions, and compute platform to assess parameter effects and capability limits. Model-based analysis was used to evaluate process contention and telemetry stream capability limits, an analysis of variance (ANOVA) was used to explore compute platform sensitivities, and first-order parameter effects were determined empirically. Model-based results indicate capability limits imposed by architecture and ANOVA results indicate algorithmic or implementation limits on performance. Empirical results from trajectory prediction tests established parameter ranges. SixSigma DoE and Simplex were explored as trajectory prediction tuning processes.

INTRODUCTION

DARPA's Strategic Computing Initiative in the mid-1980s included the development of autonomous aerial and terrestrial capabilities to improve safety and reduce logistics costs, but for controlled environments, which is the case for much of today's autonomous systems. Technology maturation typically progresses to increasingly complex environments (e.g., DARPA's Grand and Urban Challenges) but autonomy in highly contentious and adversarial environments remains particularly challenging. In additional to the environment, other challenges are regulatory and technical; the former involves meeting safety and verification requirements for non-deterministic or AI-enabled systems while the latter includes compounding complexity, performance requirements, and economies of scale.

For the better part of the last decade, NASA's Armstrong Flight Research Center (AFRC) has been developing an all-domain autonomy framework called the Expandable Variable-Autonomy Architecture (EVAA). Whereas the F-16's AutoGCAS system superseded control to command a recovery maneuver long enough to avoid a ground collision, EVAA extended that concept to multiple hazards, maneuvers, and vehicles. The framework essentially inverts the design and assurance paradigm from the specification of everything a system will do in all circumstances to the specification of what it won't do in any circumstance.

NASA AFRC was also driven by the need to enhance the civil certification process for airborne software and autonomous systems in particular. The primary challenges facing the FAA include a fundamentally serial certification process, increasing work volumes, and an exacerbating economic and engineering environment seeking to commercialize novel systems. NASA imagined EVAA as a certified assurance framework containing a conops-driven set of certified monitors that would

[1]Modern Technology Solutions, Inc., 5285 Shawnee Road, Suite 400, Alexandria, VA 22312.

reduce the certification burden to a particular integration. This also affords the rapid assimilation, sand-boxing, and fielding of complex or AI-enabled capabilities that would otherwise require significant effort to certify and field conventionally.

EVAA can be used to operate in uncertain conditions where hazards need to be mitigated to an acceptable level of risk. It can also be used for all domain operations (submersible, naval, terrestrial, aerial, and space) provided the system has an autopilot. Ensuring vehicle stability and bounded command targets is delegated to the autopilot, which thereby provides an added layer of safety.

The general trend in defense aerospace over the past decade has been towards open architectures and reusable and interoperable software components. EVAA is a government owned plug-in architecture that facilitates the reuse of certifiable software components. Users select and allocate capabilities specific to their concept of operations, vehicles, payloads, and required behaviors.

Hazard monitors bound the behavior of non-deterministic or complex systems, which in effect ensures system behavior is predictable. EVAA continually evaluates ownship and environmental states, continually assesses the viability of predefined escape maneuvers, and evaluates the consequence of not performing a maneuver. EVAA's supervisory control mechanism is comprised of a set of certifiable reversionary controllers whose commands are adjudicated by an arbiter that decides which controller's command should be employed at any moment based on relative consequence values. This allows EVAA to avoid any compounded or sequence of hazards. Each reversionary controller implements a maneuver informed by best practices that is expected to deliver the vehicle from harm after which control is relinquished to the pilot or whatever guidance and navigation system (GNC) was previously in control. The execution of a predefined recovery maneuver is essentially how ACAS-Xu, AutoGCAS, and TCAS systems work.

The automation of recoveries is a means for achieving autonomy, managing complexity, and scale. It avoids compounding complexity by both arbitrating among independent algorithms (e.g., terrain-aware air-collision avoidance) and by mediating communication through a common database, which is also known as the blackboard agent architecture first proposed by Carnegie Mellon University in 1979.[1] EVAA supports variable-autonomy in that it can be used on manned and unmanned vehicles and can be configured to override the pilot, be overridden by a pilot, or be used to drive instrumentation to improve situational awareness.

Although EVAA can be licensed through NASA AFRC's Tech Transfer Office,[2] it is still a work in progress and there are gaps in documentation that may impede adoption. Specifically, there is neither documentation on what effect its parameters have on behavior nor how to tune them, both of which are also required to understand limits on applicability. It is unknown whether a system intended for low probability hazards and Part 23 aircraft is suitable in tactical or contentious environments. Understanding the limits of this technology informs which missions, vehicles, autonomous behaviors, computing platforms, and business pursuits are suitable.

This paper begins to fill this knowledge gap and is organized as follows. The Background section reviews supervisory control in general and EVAA in particular. The four subsequent sections start by introducing additional detail regarding the subsystem under test, a brief literature review to contextualize and inform the experiment, a description of the experiment or analysis, results, and discussion. The Process Contention section explores limits on the complexity of the operational environment by estimating the maximum number of plugins EVAA can support. The Compute Platform section assesses EVAA's sensitivity to compute platforms, which informs platform selection. The Telemetry Stream section evaluates the limits on EVAA's internal telemetry stream, which can be used for real-time monitoring or control, e.g., a flight termination system that keeps missiles within a test range, or be used for vehicle to vehicle communication to coordinate swarming behavior, for example. Lastly, the Trajectory Prediction section evaluates the effects of EVAA's prediction parameters.

---

[2]https://technology-afrc.ndc.nasa.gov/featurestory/collision-avoidance

BACKGROUND

Supervisory Control

Design patterns for running multiple controllers in parallel differ in arbitration strategy and purpose. In designs where optimality is paramount, experimental controllers are weighted as co-equal or greater than a baseline controller while designs that favor fault-tolerance allow the baseline controller to either constrain or supersede experimental controllers. The simplex architecture ran baseline, safety, and experimental controllers in parallel while an arbiter decided which was in command.[2] The sand-boxing architecture proposed giving an experimental controller precedence, but constrained its commands with a baseline controller.[3]

The runtime assurance architecture (RTA) gave the baseline controller exclusivity over the plant but ran experimental controllers in parallel to record its outputs.[4] Generic RTA proposed running multiple certified recovery controllers in parallel to constrain the behavior of an experimental controller and is there a hybrid design that balances fault-tolerance and optimality.[5] RTA is intended to prevent a perfectly controllable vehicle from executing an unsafe outer-loop control action by overriding it with an alternative command from a trusted controller.

The airworthiness design standard for ensuring autonomous systems maintain an acceptable level of safety and mission assurance for routine operations is ASTM F3269-17 titled Standard Practice for Methods to Safely Bound Flight Behavior of Unmanned Aircraft Systems Containing Complex Functions, which was informed by NASA AFRC's work on RTA.[6] Multi-Monitor Run Time Assurance (MM-RTA) is an extension of RTA that includes a set of monitors while RTA-Network further extends the concept to contingencies.[7,8]

EVAA System Overview

EVAA is NASA AFRC's implementation of MM-RTA. Conceptually, EVAA is similar to an operating system in that it provides an environment for running other applications called plug-ins. EVAA is an asynchronous multi-threaded system written in C++, compiled using the GCC tool-chain, and can run on commodity hardware and any Linux distribution that supports PREEMPT-RT or priority based scheduling.

EVAA contains several built-in services that are collectively referred to as EVAA-CORE and includes a decision module ("Moral Compass"), a key-value store (Common Value Table or "CVT"), a client-server data distribution service ("CVT Stream"), multiple data validation systems, a mapping system, and a facility for inter-process communication between monitors ("Crosstalk").

The five types of plugins are hazard monitors ("monitors", "controller"), common services ("tasks"), sensor-couplers, command-couplers, and trajectory predictions. Sensor-couplers take observations from some sensor, optionally digest or filter that information, and write to the CVT. Monitors are vehicle controllers that perform a specific threat assessment and communicate recovery commands to the Moral Compass as warranted. Monitors may be reactive with short cycle times for safety-essential behavior or deliberative with longer cycle times for strategic and mission behaviors. In the event multiple monitors simultaneously issue recovery commands, the commands are reconciled by EVAA's arbiter ("Moral Compass"), which decides which controller will supersede control of the vehicle at any given moment. The Moral Compass forwards the selected command to the command-coupler, which translates EVAA's internal vehicle commands into commands and protocols understood by the vehicle's autopilot.

NASA released three versions of EVAA during the Resilient Autonomy (RA) Joint Concept Technology Demonstration (JCTD) project between 10/2018 and 08/2021 as listed in Table 1. Note that of the 17k software lines of code (SLOC) in the baseline version of EVAA prior to RA, 7k SLOC was for EVAA's internal mapping system. The first RA release included major modifications to EVAA-CORE and enhancements to the terrain avoidance monitor. The air collision avoidance suite added 36k SLOC to RA-2.3. Version RA-3.3 added an additional 10k SLOC for well-clear, 1.5k SLOC for mission and contingency behaviors, and 5k SLOC to EVAA-CORE. Although the

Table 1: EVAA Release History

Rel.	Ver.	MMYY	SLOC (k) Total	CORE	Behavior
	Pre-RA	05/18	17	12	5
1	RA 0.8	09/20	25	21	4
2	RA 2.3	04/21	62	21	41
3	RA 3.3	08/21	80	26	54

version used in this work was RA-0.8, the sensitivities evaluated in this work regard limits dictated by architecture, so the addition of new functionality or slight modifications to existing functionality is expected to have negligible effect on outcomes.

PROCESS CONTENTION

Introduction

The "E" in EVAA stands for Expandable and refers to the nonfunctional quality of extensibility. EVAA's architecture allows it to assimilate an arbitrary number of plug-ins. The concept of extensibility is implied in most supervisory control architectures discussed in the Background section. The RTA standard[6] is explicit about a multiplicity of escape trajectories, the MM-RTA paper[7] is explicit about a multiplicity of hazard monitors, and the RTA-Network paper[8] incorporates contingencies, which further increases the number of plugins.

The number of plugins required for a particular operation depends on the complexity of the environment. Hazards can be external or internal (e.g., degraded control) and static (e.g., territorial geo-fences) or dynamic (e.g., other vehicles, inclement weather). Each hazard monitor is informed by one or more sensor-couplers and each monitor may use a multiplicity of escape trajectory plugins.

These are typical examples of plugins used to ensure safety, but since EVAA was intended to bound the behavior AI-enabled systems to operate in increasingly complex environments, it is very likely the number of AI-enabled perception, navigation, and planning plugins will outstrip the number of safety plugins. NASA explored contention between EVAA and external processes on the same host by running a `while(true)` loop and confirmed no impact on EVAA's ability to schedule processes. However, they have not explored internal contention, i.e., limits on the number of plugins that EVAA can manage.

Literature Review

The author in (Reference 9) subdivides performance evaluation into prediction and measurement concerns that map to the design and maintenance parts of the software life-cycle, respectively. Predictive evaluation uses abstract models of a proposed system to assess whether its theoretical performance can meet quality of service guarantees, whereas performance measurement uses runtime analysis to characterize an implemented system's actual performance.[9] Note that architectural problems are attributed to anti-patterns that can not be resolved by tuning.[10]

Models are often defined in terms of specialized versions of UML that allow the specification of timing, synchronized access to shared resources, control flows, internal state, and parameter dependencies.[9] Different types of UML diagrams capture unique aspects of a system's performance, e.g., deployment diagrams specify platform capabilities, use case diagrams specify workload, and sequence diagrams capture control flows.[10] Although noting the broader acceptance of model-driven engineering and the increasing maturity of performance engineering tools over the past decade, researchers identified the two principal barriers to adoption for component based software modeling: the availability of usage profiles and model expressiveness.[9]

Components are the compiled and reusable units of aggregation and (Reference 9) provides a broad survey of performance evaluation solutions for component based systems that incorporate classical probabilistic or stochastic performance modeling techniques. A component's performance is a function of its usage profile, contention over shared or synchronized resources, other services, and the platform.[9] The problem is that the usage profile is required to accurately model and predict a system's performance[9,10] but component designers can not anticipate every possible use case or usage profile and integrators only acquire a usage profile as their specific system progresses through the development life-cycle.[9]

This paragraph provides a brief overview of a select set of performance prediction and evaluation tools described in (Reference 9) and adopts the author's convention for enclosing system names in parentheses. (CB-SPE) uses the industry informed RT UML Profile for modeling and an LQN (layered queuing network) solver to predict a real-time system's performance.[11] (PECT) provides a language for specifying process rates and demands to analyze the properties of real-time embedded systems and uses either MAST simulation or RMA (rate monotonic analysis), which is used to predict the schedulability of real time tasks.[9] (ROBOCOP) provides a basic parametric model that is translated into a task tree, which can then simulate different scheduling policies.[9] (PALLADIO) uses control flow graphs attributed with resource demands that are translated into EQN or LQN queuing networks and solved either by simulation or numerical analysis.[9] (NICTA) uses an annotated activity diagram that is translated into a queuing network and solved using MVA (mean value analysis).[9]

For implemented systems, (COMAERA) subdivides software into computational atoms used to derive a platform-agnostic measure of utilization, which can then be used to predict the performance of a bench-marked target.[9] The authors in (Reference 12) proposed either translating source-code into an instrumented virtual instruction set or translating assembly code to isolate the uncontrolled effects of compiler optimizations. Their work estimated utilization within 10% of actual for two of three tasks, but significantly overestimated another.

General strategies for evaluating performance include the use of instrumentation libraries, timing discrete computational steps, timing distinct control flows, or by taking a component-centric approach that uses regression to derive demand functions.[9] The authors in (Reference 13) proposed transforming software models into performance models throughout the development life-cycle using an intermediate model specified using the UML MARTE Profile.[10] Others used UML process algebras to generate state spaces or models from which performance could be estimated using a stochastic Markov process.[10]

All of these tools and techniques assume the parameters of the usage profile are constant, but (Reference 14) contemplated the effect of variable usage profile parameters and the identification of parameter boundaries where the overall system would fail to meet its performance requirements. Earlier those researchers proposed a method for finding bottlenecks by performing a sensitivity analysis on each parameter using Pareto optimization, where the objective was for the system to remain feasible and where the Parto front identified the boundary between feasible and infeasible values. Because one-dimensional analysis assumes all parameters are independent, they both incorporated and extended that work to a pairwise two-dimensional sensitivity analysis for finding the maximum allowable variation of a system's usage profile.[14] Note that the novelty of their approach is varying usage profile properties rather than system parameters for tuning a system. Their method is to perform a depth-limited recursive binary search of a two-dimensional parameter space to identify the 2D points where the system becomes infeasible. The output is visualized as pairs of linear plots with roughly constant slope until the system becomes infeasible at which point the slope changes significantly.

Experimentation

EVAA can load as many plug-ins as the underlying hardware can support. The ability to load plug-ins is a function of space, but the ability to run plug-ins is a function of time. Scheduability is far more constraining than storage or memory capacity. Hardware parallelization using multiple processors (SMP), multiple cores (CMP), or multiple threads (CMT) support scaling the number of simultaneous processes, but ultimately interleave and serialize workloads on each computational unit. Operating systems offer a variety of scheduling policies for deciding which process runs at any given moment, but regardless of policy, only one process can run at a time since the CPU can only perform one fetch, increment, decode, execute cycle at a time.[15]

The exact number of plug-ins that EVAA can support on a specific computing platform could be determined experimentally by evaluating performance for a particular assignment to parameters, but an exact solution for a particular platform has limited value and would require an impractical number of tests. From the literature review, the customary approach to estimating performance involves a simulation of a simplified system. However, since model-based approaches were criticized for low fidelity, a purely analytic model appeared sufficient for assessing system limits.

EVAA's process related parameters - affinity, period, priority, and number of processes - are positive integers, which means the parameter state space is inherently discretized, finite, and combinatorial. Affinity is a 1-byte bit-field where each bit identifies the core(s) a process can run on and is therefore processor dependent, but since it is defined as a byte, the range is $\mathbf{A} = \{a \in \mathbb{Z}^+ | 1 \leq a \leq 8\}$. Period is defined as a 4-byte integer, but since a negative time period is impossible and the units are milliseconds, let the range be $\mathbf{T}_0 = \{t \in \mathbb{Z}^+ | 0 < t \leq 1000\}$. The range for priority is $\mathbf{P} = \{p \in \mathbb{Z}^+ | 0 \leq p < 100\}$. For a set of hardware platforms $\mathbf{H}$, K user plug-ins, EVAA-CORE's three internal scheduled processes, the entire parameter space is defined by $\prod_{i=1}^{K+3} \mathbf{A} \times \mathbf{T}_0 \times \mathbf{P} \times \mathbf{H}$.

That model can be simplified as follows. A single computational unit, or core, captures the schedulability problem and can be extrapolated to multiple computational units, so the CPU-core affinity parameter is dropped from the model. To reduce the size of the model's parameter space, Period was constrained to increments of 100ms: $\mathbf{T} = \{t \in \mathbb{Z}^+ | 100t \leq \tau\}$, where the overarching MoralCompass period $\tau = 1000$. The increment and choice of τ limit the number of processes per core to 10. Here the model assumes all processes run once per τ, which allows priority to be dropped. Processes that run more than once per τ can be modeled as aggregates.

Hardware effects execution speed, so the set of hardware configurations is absorbed by the period parameter. Because the operating system requires time to switch between processes, a CPU utilization threshold is defined in terms of the percentage of total processor time as $\mathbf{U} = \{u \in \mathbb{Z}^+ | 50 \leq 10u \leq 100\}$. Finally, to prevent EVAA-CORE's three internal processes from being counted on each core in deployments with C cores, let $k = K + round(\frac{3}{C})$.

The model only requires that the sum of periods for an arbitrary number of processes is less than or equal to the target utilization. The model is defined by Equation 1. Algorithm 1 lists the procedure for generating its entire state space, which contains $2E^3$ configurations.

$$\sum_{j=1}^{k} \left(\prod_{i=1}^{k} \mathbf{T}_i \right)_j \leq u \in \mathbf{U} \tag{1}$$

Algorithm 1 To generate the model's entire state space **S**, this algorithm takes the cross product of periods for all processes where the sum is less than or equal to the utilization threshold. Here, **T** is a set of process periods, C is the number of processor cores, k is the number of processes per core that takes into account EVAA-CORE's 3 internal processes, **S** is the state space, and **U** is a set of utilization thresholds.

1: $\mathbf{T} \leftarrow \{t \in \mathbb{Z}^+ | 100t \leq \tau\}$
2: $k \leftarrow round(K + \frac{3}{C})$
3: $\mathbf{S} \leftarrow \mathbf{T}_1 \times \ldots \times \mathbf{T}_k$
4: $\mathbf{S}_u \subset \mathbf{S} \leftarrow \{\mathbf{s} | \sum_{i=1}^{k} s_i \leq u \in \mathbf{U}\}$

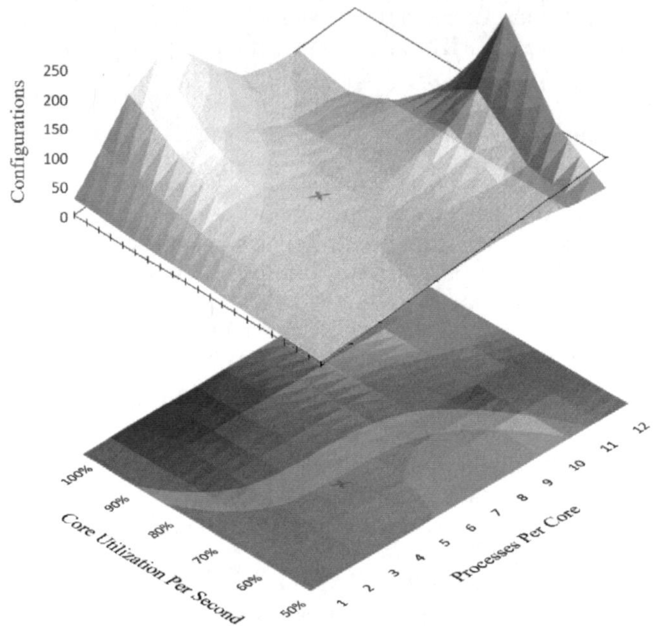

Figure 1: Process Schedulability and Contention Model

Results

Figure 1 depicts a number of salient features of the model's state space. The 3D surface represents the number of scheduling options, which is a function of the utilization threshold and number of processes. EVAA's capability to support a variety of scheduling options is extremely valuable from an integration point of view. Real-life deployments on larger UAS (e.g., MQ-1/ER) involve multi-tenant systems running on shared resources and configurability allows EVAA to allocate processes across distinct compute resources, whether physical or virtual. Configuration options are positively correlated to utilization and inversely correlated to the number of processes.

Any 2D plane along the utilization and configuration axes has a Gaussian curve, but a detail not shown on the figure is that in any particular configuration the set of periods has a Poisson distribution, i.e., Pareto-like. That means a relatively few dominant processes will take a larger proportion of computational time.

The larger configuration space for high utilization thresholds is due to combinations and permutations of parameters that are equally viable. Note that this model unexpectedly captured the utilization upper bound according to rate-monotonic scheduling, where scheduability converges to 70% as the number of processes increases to infinity. That feature is visible on the far right of the surface plot and was colored gray to denote an infeasible region.

The surface on Figure 1 was projected onto a 2D topological map to emphasize the feasibility of regions within the model's parameter space. Although the utilization threshold is typically 70%, it can be as high as 85% when there is only one scheduled process. The amount of operating system overhead is positively correlated to the number of scheduled processes and requires a lower utilization threshold when there are many processes. These differences in utilization at the extremes creates the curved profile shown on the topological plot. Again, since the model limits process periods to increments of 100ms and the overarching period τ to 1000ms, the region beyond ten processes is shown in gray to denote infeasibility.

According to this model, the optimal configuration that maximizes utilization, the number of processes, and configuration options occurs at the saddle point indicated by the "X" on the surface plot, which indicates an upper limit of 4 total processes. When allocated to a single core CPU this model permits only 1 user process in addition to the 3 built-in CORE processes. However, EVAA-CORE processes can be distributed across cores in a multi-core environment so a dual-core processor can accommodate an additional 5 user processes and an octa-core could support 29. A quad-core could support 13 additional user processes, which aligns with NASA's empirically determined upper limit of 16 total processes. These estimates are conservative because the model assumes that processes spend their entire time budget. In reality, slack is often built-in to each period and a portion of that slack can be exploited to run additional user processes per core, but with greater risk of adverse contention.

COMPUTING PLATFORM SENSITIVITY

Introduction

Consistent with the sub-scale work at AFRC, the team at NASA was keen to minimize the size, weight, power, and compute (SWAP-C) requirements to maximize EVAA's integrability. During the earlier Traveler program, NASA deployed EVAA on a 32-bit octa-core ODROID XU4 single board computer (SBC) on both the Bird's Eye View FireFly6 VTOL and a custom built fixed-wing aircraft.

During the Resilient Autonomy (RA) Joint Concept Technology Demonstration (JCTD) with the FAA, industry vendors, and DoD stakeholders, NASA was further constrained by the compute platform's country of origin. For the JCTD – utilizing L3's FVR-90 and FVR-10 VTOL aircraft as well as a manned SR-22 – NASA selected a hardware suite comprised of one quad-core Up^2 processor for ten EVAA processes and a pair of NVIDIA Jetsons for six external computer vision (CV) and terrain referenced navigation (TRN) processes.

NASA's other platform selection criteria included properties such as form factor, environmentals, power, clock-speed, number of cores, RAM, connectivity, cost, compiler support, and O/S support. The gap that this paper attempts to fill regarding platform selection is the effect various hardware technologies have on cycle time.

Literature Review

To demonstrate how safety-critical systems can be impacted by the choice of hardware, researchers in (Reference 16) ran five workloads on MCU and FPGA platforms and discovered differences in reliability. Specifically, they found that only MCUs were susceptible to measurement error on a counting task. While both platforms were impacted by the timing and multiplicity of CAN-bus messages, FPGAs had twice as many errors. The authors included verifiability as platform selection criteria and noted that FPGAs may be easier to verify since they lack interrupts.

In (Reference 17), faults were injected into a specific workload running on two different CPU and O/S combinations to reveal each platform's sensitivity to a variety of errors. The research indicated that the Pentium 4's variable instruction and data widths that improve memory management also led to much greater sensitivity to error. This difference was traced back to the operating system's support for the Pentium 4 processor, which did not define an interrupt for stack overflows, so the error tended to propagate until hitting another error condition. The PowerPC G4's fixed-width instructions and data addresses had three times fewer memory pointer errors, but its larger number of general purpose registers meant potentially adverse data could persist in memory much longer than in the Pentium.

Rather than use an exhaustive approach and test specific hardware combinations, (Reference 18) proposed a method using constraint satisfaction and heuristics. Recognizing the large number of possible configurations, (Reference 19) proposed parameterizing hardware properties that could be programmatically manipulated and optimized given a specific system and workload.

The key contribution in (Reference 20) was the development of a statistical framework for predicting system performance that was neither specific to any specific computing platform nor required system modeling but rather focused on cycle time variations in maxima, i.e., in terms

Table 2: Listing of Hardware Variations

Factor	Levels	Variants
Disk_type	2	eMMC, SD
Disk_speed_Mbs	3	10, 50, 90
RAM_GB	3	1, 2, 8
RAM_channels	2	single, dual
RAM_speed_Gbs	3	8.5, 14.9, 38.4
RAM_type	3	DDR3L, LPDDR3e, LPDDR4
Process_nm	3	14, 28, 65
Arch	2	ARM, x86-64
ISA	2	RISC, CISC
Cores	3	1, 2, 4
Clock_speed_Ghz	11	0.2, 0.3, 0.4, 0.7, 1.0, 1.1, 1.2, 1.4, 1.6, 1.8, 2.0
Cache	4	2MB L2, 32k L1 + 512k L2, 64k L1 + 1MB L2, 64k L1 + 4MB L2

Table 3: Computing Platforms

SBC	CPU	Memory	Storage	O/S
Up^2	Quad-core Pentium N4200 x86-64 CISC, 2.5 GHz, 2MB L2, 14 nm die	8 GB dual channel LPDDR 38.4 Gbs	64 GB SanDisk DDR50 eMMC, 50 Mbs	Yocto Sumo PREEMPT-RT
Odroid XU4	(a) Quad-core 32-bit ARM Cortex-A15 RISC, 2 GHz, 64k L1 + 4MB L2, 28 nm die (b) Quad-core 32-bit ARM Cortex-A7 RISC, 1.4 GHz, 64k L1 + 1MB L2, 28 nm die	2 GB dual channel LPDDR3e 14.9 Gbs	(a) 64 GB eMMC 5.0, 90 Mbs (b) 16 GB SD Card 3.0, 10 Mbs	Ubuntu 20.04 PREEMPT-RT
Parallala 16	(a) Dual-core 32-bit ARM Cortex-A9 RISC, 1 GHz, 32k L1 + 512k L2, 65 nm die (b) 16-core Zync Z7020 (unused)	1 GB single channel DDR3L 8.5 Gbs	8 GB SD Card, 10 Mbs	Ubuntu 18.04 PREEMPT-RT

of worst-case execution times (WCET). The authors noted cache conflicts, varying loads across compute resource, power management, and O/S jitter as sources of cycle time variation for data intensive, synchronized, and large scale applications. Their framework included both parametric and non-parametric models that were compared to measurements from two high performance computing platforms running six different workloads and found, in pertinent part, that model accuracy varied with workload.

Experimentation

Although Table 2 summarizes all hardware variations in this study, the actual number of configurations under test were limited by the single board computers (SBC) under test, which are listed in Table 3. These platforms primarily differed in terms of core count, clock speed, storage technology, and RAM, but also varied in terms of architecture, instruction set, die size, cache, RAM technology, RAM speed, and disk speed.

The ODROID XU4 combines two ARM Cortex CPUs with different cache sizes and clock-speeds and the `cpufreq` utility was used to throttle the clock-speed on each bank of cores. The ODROID also permitted tests using both SD Card and eMMC storage technologies, the latter of which had a different data rate than the Up^2's eMMC.

Each platform was tested under the same workload using EVAA configured with NASA's improved ground collision and avoidance system (iGCAS v0.8), whose sensor is a mapping system that uses map-file parameter data items (PDI), and that uses the GA Prediction plugin. The primary metric for the iGCAS workload was cycle time in milliseconds, which was captured by wrapping each iGCAS call within EVAA's Timer helper class. Cycle times were saved to a file using EVAA's

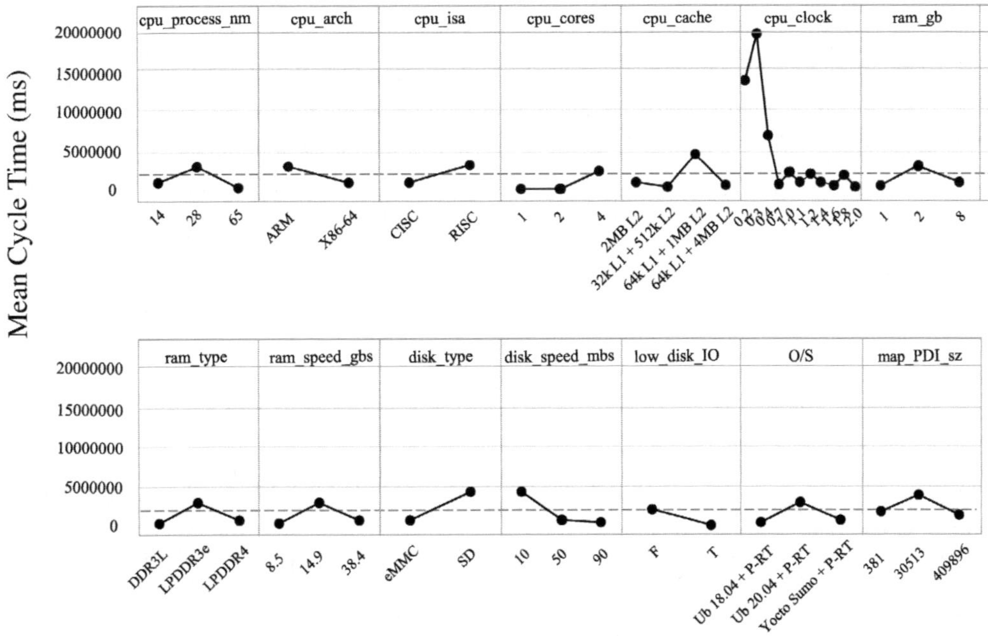

Figure 2: ANOVA: Main Effects for iGCAS Workload

built-in Logger. Each configuration was run once and each run involved a bash script that ran `vmstat` in parallel to capture the unixtimestamp and CPU and memory usage, which were secondary metrics. The unixtimestamp was used to correlate iGCAS run times from EVAA logs to resource utilization from `vmstat` and plotted to explore jitter and frame overruns. Once correlated, the data was loaded in SQL Server where it was aggregated, encoded, and formatted.

Since the hardware variations in Table 2 have categorical inputs while cycle time, CPU, and RAM utilitization have continuous outputs, the dataset lent itself to evaluation by an analysis of variance (ANOVA). ANOVA assesses whether there are statistically significant differences between means across samples and whether there are relationships between input factors.[21] MiniTab 20.2 was used to perform the ANOVA. However, MiniTab requires a large number of combinatorial tests to estimate statistics so the quantitative general linear model could not be used. Instead, the qualitative main effects plots were produced, which illustrate the relative strength of factors and their influence on response variables. Main effects are typically interpreted in conjunction with interaction plots that illustrate dependencies between factors, but those too require a large number of combinatorial tests. The only combinations of factors tested in this study were: arch v cores, ISA v cores, cores v cache, cores v GHz, cores v RAM_GB, cores v RAM_type and RAM_speed_Gbs, cores v disk_type and disk_speed_Mbs, cache v disk_type and disk_speed_Mbs, and GHz v disk_type and disk_speed_Mbs.

Results

As can be seen from Figure 2, clock speed has the greatest impact on iGCAS cycle times but only to a certain point, which indicates algorithmic limits on performance. A far second factor with the greatest impact on cycle time was disk type, which is likely due to the storage technology's bus speed and underlying physical layer protocol, which may be consuming clock cycles. Note that workload cycle times bottomed out with disk speeds over 50 Mb/s. Next in significance was CPU architecture with x86-64/CISC outperforming ARM/RISC, which appear similar to the findings in (Reference 17).

Although main effects plots for resource utilization are not shown, disk type and CPU architecture had the greatest effect on CPU utilization with eMMC and x86-64 outperforming. RAM type, speed, and capacity had the greatest impact on memory utilization with more recent RAM technologies logarithmically outperforming. Finally, dual channel RAM outperformed single channel memory with respect to memory utilization, but single channel RAM edged out dual channel RAM in terms of workload cycle time.

That the impact of clock speed on cycle time bottomed out after 700 MHz points to limitations imposed by the algorithm and warrants reconsideration of how reusable any plugin may be. The workload was profiled using `perf` and EVAA accounted for $35.3\% \pm 2.0\%$ of CPU usage, the GA prediction accounted for $6.3\% \pm 1.6\%$, and iGCAS monitor accounted for $0.6\% \pm 0.2\%$ CPU usage. From this it appears that workloads may need to be coupled to vehicle performance classes.

TELEMETRY STREAM SENSITIVITY

Introduction

EVAA has a built-in key-value store called the Common Value Table (CVT) as well as a TCP server (Intent Manager, a.k.a. CVT Stream) that streams data from that database. NASA exploited this during flight test to drive a ground control station (GCS) heads up display (HUD) and to remotely update CVT data such as mission, run number, or to set flags to trigger maneuvers. The CVT Stream subsystem packetizes data and serially distributes packets to any connected clients. The CVT Stream's configuration file allows users to define the packet size, maximum number of clients, and send rates for CVT and ownship data. This part of the paper assesses the performance limits of the CVT Stream.

Literature Review

A data distribution system is comprised of hardware, communication mediums, device drivers, protocols, and communication architectures. Researchers in (Reference 22) demonstrated a low-cost microsecond-scale testbed involving a software defined radio to evaluate the performance of wireless hardware under varying power levels or in the presence of congestion and interference and traced device behavior down to the underlying protocols, chipsets, and hardware interfaces. They also studied capture effects, which regards device behavior when multiple signals are received at the same time, and found that although the capture effect requires a minimum SIR and delay, it can only occur if the delay is sufficiently small. That means receivers lock onto a packet and block until the first transmission completes. The differences in observed behavior were attributed to each vendor's implementation of the 802.11 technical standard and could be used for device fingerprinting. In (Reference 23), differences in performance were traced to device drivers that may be performing extraneous copies.

The authors in (Reference 24) reviewed legacy object-oriented communication architectures (e.g., CORBA, DCOM) that comingled data and logic, which led to tightly coupled integrations, while point-to-point architectures (e.g., SOA) required explicit endpoint identification. Those were contrasted to the concept of a global data space in the publish/subscribe ("pub/sub") architecture that mediates messaging between a dynamic set of systems at data rates commensurate with each client's needs. Mediated communication both simplifies the addition, removal, or substitution of systems and can reduce the number of point-to-point connections within an integration. Pub/sub systems are event-driven: clients register subscriptions to select topics and receive topic specific messages only when publishers post new information.

The researchers in (Reference 25) noted pub/sub architectures exhibited a heavy tailed zipf distribution where a small minority held a large number of subscriptions and a vast majority of topics had a few. They noted that although pub/sub anonymizes sources, clients were still coupled to specific sources. They argued topic handles were symbolic in nature and proposed a semantic variation of the pub/sub architecture that decouples subscribers and publishers, which may be better suited where topics and sources are highly dynamic or where centralization presents an unacceptable risk from a single point of failure.[25]

The User Datagram Protocol (UDP) is an unreliable connectionless transport protocol for exchanging messages with minimal overhead while the Transmission Control Protocol (TCP) is a reliable connection-oriented point-to-point byte-streaming transport protocol suitable for large file transfers.[26] The widespread use of these protocols is ingrained in networking equipment, operating systems, and API such that there are significant barriers to entry for novel protocols.[27] A connection establishes a specific network path and although data link layer protocols ensure reliability for each hop, TCP ensures reliability for a network path using acknowledgments and sequence numbers that can detect lost, corrupted, duplicated, or out-of-order packets.[26] TCP implements congestion control to adapt transmission speeds to available bandwidth .[23]

As noted in (Reference 28), TCP was designed for reliable delivery not timely delivery. Packet retransmission, increased congestion, and greater latency result from either inadequately sized buffers on the receiver or the TCP sender increasing its transmission speed until packet-loss signals channel capacity saturation, which are known as the incast and bufferbloat problems, respectively.[27] TCP does not report intermittent packet losses, but rather only permanent losses that close the connection.[28] TCP's congestion control mechanism can reduce the senders transmission speed until packets timeout and require retransmission and thereby exacerbate the congestion problem.[27] TCP is slow to recover from channel quality fluctuations typical of wireless links and the in-order guarantee can block other data flows over a shared TCP connection, which is known as the head of line (HoL) problem.[27] TCP is CPU intensive and can saturate a modern CPU core when transmitting at 10Gbps and has an effective rate of 13Gbps on a 40Gbps host.[23]

The Datagram Congestion Control Protocol (DCCP) adds congestion control to UDP over two half-duplex connections, but neither guarantees reliability, order, nor retransmits lost packets.[27] A reliable connection-oriented alternative that utilizes bandwidth better than TCP on optical or high bandwidth-delay product (BDP) networks is the UDP-based Data Transfer (UDT) protocol,[29] but (Reference 23) notes that protocol overhead limits its use for high speed networking applications. The authors in (Reference 28) proposed a real-time messaging protocol, the Delay Bounded Reliable Messaging Scheme (DB-RMS), for bounding latencies using configurable recovery policies and timeout parameters. Although promising, there doesn't appear to have been any follow-on work.

Remote Direct Memory Access (RDMA) is a zero-copy protocol that allows applications to register buffers on network adapters and thereby bypass O/S paging, interrupts, and copying between user and kernel spaces, but requires a virtually loss free channel.[23] Google's Quick UDP Internet Connection (QUIC) protocol operates entirely in user space and mitigates the head-of-line problem with multistreaming, which isolates blocking due to packet loss on another stream.[27] However, QUIC's CPU utilization is almost four times higher than TCP, frequent library updates may lead to incompatibilities, and QUIC also suffers from packet reordering over wireless links.[27] The Stream Control Transmission Protocol (SCTP) uses multistreaming to improve throughout and recovery on wireless networks and also uses multihoming, which tracks the quality of primary and backup routes, but may be unable to route packets fragmented by the Internet Protocol (IP) if the maximum transmission unit (MTU) changes.[27]

Distinct sequences of message exchanges in any communication system or protocol can be generically modeled to estimate traffic using finite state machines (FSM) and optionally enhanced by cost functions that take into account context dependent performance factors.[30] Performance metrics are fundamentally time or space centric. Data volume in terms of size and message quantity was used in (Reference 30) while (Reference 31) defined performance metrics in terms of response delay and signaling or protocol overhead and (Reference 24) defined QoS in terms of deadlines and latency-budgets. With respect to instrumentation, `netperf`, which splices file descriptors to avoid copying buffers, was used to benchmark TCP, `nuttcp` was used to benchmark UDP, and `perf` has been used to profile an application's memory utilization.[23] `Tcpdump` was used to capture data in (Reference 22).

Table 4: Configurations for Estimating R

Data Type	CVT_Fields	RX_Packets	Period (s)	Bytes/s
uint8	100	100	2.219E+4	1.903E+3
float	100	100	2.486E+4	2.089E+3
uint64	100	100	2.916E+4	2.514E+3
uint64	100	10000	2.856E+6	3.221E+3
uint8	100	10000	2.244E+6	2.565E+3

Experimentation

Rather than benchmark the performance of the CVT Stream on specific hardware, here we take a model-based approach and assess performance limits with respect to parameters. Packet size is given by the minimum transmission unit parameter (mTU), a cap on the number of clients is given by maxClients, and delays for streaming CVT and ownship data are specified in milliseconds using CVT_SEND_DELAY and CORE_SEND_DELAY parameters respectively.

The mTU is defined as a C++ uint32 with range $[0, 2^{32}]$ but the effective range is $[20, 1500]$ since the packet header is 20 bytes and the maximum transmission unit for Ethernet packets is 1500 bytes. The maxClients parameter is defined as a C++ int with range $[2^{31} - 1, 2^{31}]$ but, consistent with the limits of the Piccolo and MAVLINK protocols, the effective range is $[0, 255]$. Both delays are defined as C++ int, but the effective range for each is $[1, 10^{10}]$ milliseconds, which appears to reasonably cover vehicles from hypersonics moving around 25 feet per millisecond to spacecraft periodically communicating every 100 days or so.

The system of operation involves an incrementing loop that puts its thread to sleep for a duration equal to the greatest common divisor (GCD) of CVT_SEND_DELAY and CORE_SEND_DELAY. If the increment equals some multiple of each delay period, the system packetizes and sends data to any listening clients. Ownship information is correlated and therefore transmitted as one logical structure called Core Data, which is simply packed unpadded and sent. Due to its simplicity, there is nothing more to explore regarding the Core Data stream.

The mTU parameter only applies to CVT data. To packetize CVT data, the system loops through the data elements defined in the CVT's configuration file until the packet exceeds the mTU. At that time, the packet is sent to all clients and the system resumes packetizing and sending data until all outbound parameters have been delivered. This process takes time so the question here regards the point at which the data volume exceeds the algorithm's ability to finish before the next iteration.

The effect of data volume on the mTU pack and send cycle is absorbed into the evaluation of cycle time per whole iteration, which means the system can be modeled by EQUATION 2:

$$\frac{\mathbf{V}}{R}\mathbf{C} \leq t \in \mathbf{T} \tag{2}$$

where $\mathbf{V}$ is the volume of CVT data in bytes, $\mathbf{C}$ is the number of clients, $\mathbf{T}$ is a subset of the interval $[10^{-3}, 10^{7}]$ in seconds, and R is the number of bytes that can be packetized and sent per second. The sets in the model were discretized as follows: $\mathbf{V} = \{1^{v} \mid v \in \mathbb{Z}^{+} \; and \; 0 \leq v \leq 10\}$, $\mathbf{C} = \{25c + 1 \mid c \in \mathbb{Z}^{+} \; and \; 0 \leq c \leq 11\}$, and $\mathbf{T} = \{10^{t} \mid t \in \mathbb{Z} \; and \; -3 \leq t \leq 7\}$. The decision boundary is where the cross product $\mathbf{V}/R \times \mathbf{C} \leq t \in \mathbf{T}$.

Results

To estimate R, EVAA was configured and tested as listed in Table 4, which yielded approximately 2500 bytes/s. To coarsely represent both the model's three independent variables and one dependent output variable as a voxel map, model indices were used to index voxels while voxel color on the red-hot color-scale indicates viability. That produced the voxel map shown in Figure 3, which

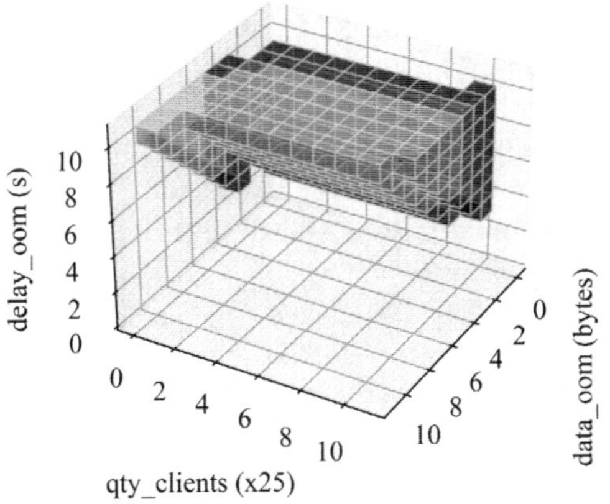

Figure 3: Viability Voxelmap

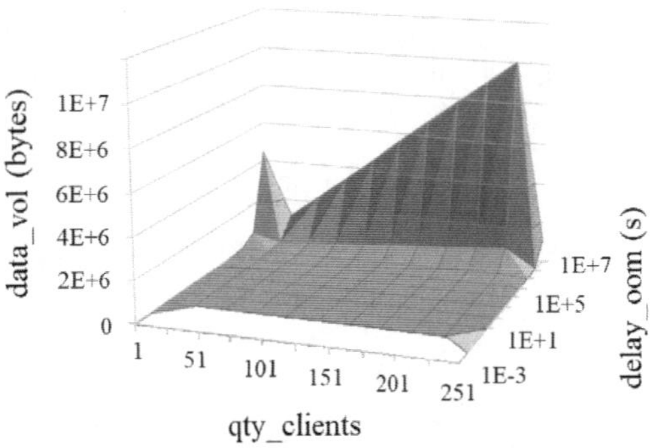

Figure 4: Throughput Behavior

shows diminishing data volume with an increasing number of clients, longer cycle times with more clients, and larger data volumes with longer cycle times. This model quantitatively estimates the Pareto-front of viable parameter combinations.

Figure 4 presents an alternative view with the data dimension now representing total throughput in terms of both CVT data and number of clients rather than just CVT data. That figure shows a very slightly inclined plateau followed by an exponential increase in bandwidth with delays longer than one second. The sawtooth profile is a result of the multiplier effect from the number of clients. Throughput increases until it hits a limit at which point a longer delay is required and the profile shifts to a larger delay.

As shown on Figure 4, the largest parameter combinations are an order of magnitude under Gigabit ethernet speeds, which can transmit up to 125 MB/s, and 80% of the bandwidth for 100 Megabit ethernet. However, commercial wireless radios used between the GCS and vehicle are typically around 25 Mbps, or around 3 MB/s. In terms of data type and volume and assuming a noiseless wireless channel, the upper bound for one client is about 625 floats/s whereas the upper bound for 250 clients is 2 floats/s. Assuming the outbound data is comprised of 15 floats for position, velocity, acceleration, attitude, and turn rates, the telemetry rate upper bound for one client is about 40 Hz and 0.17 Hz for 250 clients.

TRAJECTORY PREDICTION SENSITIVITY

Introduction

Mark Skoog, NASA's Principal Investigator behind EVAA, had earlier developed the F-16's Auto-GCAS system, but that system was tightly coupled to a single vehicle and involved one hazard and one recovery maneuver. He extended that concept to multiple hazards that had their own escape maneuvers during NASA's Traveler project and later made vehicle agnostic through architectural changes implemented during the Resilient Autonomy JCTD.

Making EVAA vehicle agnostic required decoupling EVAA's decision system from the vehicle's autopilot, which was satisfied using autopilot-specific plugins called command-couplers. Another vehicle specific dependency existed between hazard monitors and trajectory predictions, which required factoring out predictions as plugins. Note that a set of monitors can share a particular prediction and, alternatively, a particular hazard monitor can use a multiplicity of predictions.

In order to mix and match monitors, predictions, and command-couplers, prediction parameters had to be allocated to hazard monitor configuration files. This would allow, for example, geo-fence and iGCAS to share a vehicle trajectory model such that geo-fence could command a relatively gentle recovery and iGCAS could command an aggressive recovery, though both are based on the same trajectory model.

NASA developed a 15-parameter fixed-airspeed prediction model for relatively slow (e.g., $\leq$ 100-knots) fixed-wing and VTOL aircraft during the Traveler program. During Resilient Autonomy, NASA developed a 71-parameter model for thrust-limited large UAS and general aviation aircraft moving up to 300-knots. This general aviation prediction produces trajectories for accelerated turns or climbs while taking into account lateral winds and downdrafts.

Tuning these trajectory predictions is a manual, iterative, and weeks-long process involving a series of test flights using NASA's pilot activated recovery system (PARS) that commands a predefined maneuver from the ground. Data captured with varying initial conditions and command targets are run through an unknown tuning process that aligns the model's prediction with the vehicle's actual performance. The process of estimating parameters using that data is not publicly available.

Literature Review

To reduce the manual, iterative, and lengthy process of tuning parameters, researchers in (Reference 32) implemented a multi-step sensitivity analysis (SA) for a 100+ parameter complex microscopic traffic simulator. Moreover, researcher employed a five-step methodology to group parameters based upon their sub models, develop a mapping function to relate a combination of parameters within the same group based upon each parameter's specified range, and lastly apply variance-based SA to further simplify the parameters into a final set before the completion of the tuning process utilizing a Kriging-based calibration. The research indicated that the multi-step SA made it possible to reduce potential model evaluations up to as much as 80%; however, it notes that the process needs to be further refined due to lack of testing and evaluation for the entire selection and tuning process.

In (Reference 33), quasi-optimized trajectory-based elements effect (quassi-OTEE) and Kriging-based SA were utilized in a multi-step process to determine discardable parameters before computing first-order indexes for determination of important parameters' ranks. Specifically, the research indicated that there were limitations between both approaches requiring them to be utilized in a multi-step process to reduce the 192-parameter model to just 14 important parameters that were being tested in the experimental studies. Note that the results were developed utilizing a preset range for each of the important parameters and uniform distribution curve, while demonstrating the limitation of each approach with no one approach being 100% accurate for important parameter detection.

As an alternative to ranking important parameters, (Reference 34) proposed the use of factor fixing, also known as fixing the value of a parameter based upon the parameter's effect on the output uncertainty before completing the calibration process. Utilizing a variance-based SA, parameters with a value of less than 2% on the total sensitivity index were considered non-influential, allowing researchers to apply factor fixing to those specified parameters. Using the factor fixed parameters, a reduced model was developed and tuning procedures against all trajectories were conducted comparing the reduced model and full model. The results suggest that there are very slight deviations between the estimation of the parameters; however, the charts demonstrate very similar shapes and values.

The key contribution in (Reference 35) and (Reference 36) was the focus upon parameter tuning for a specific maneuver regarding an automated vehicle and underwater glider. Both studies implement a version of a sensitivity analysis and produce results indicating that specific groups of parameters have a more significant impact on the specific motions required to complete the maneuver, which indicates that certain parameters are considered non-influential like the results in (Reference 34).

Experimentation

For the general aviation prediction, a full factorial study using only three values for each of the 71 parameters would require 7^{33} tests, which at 5 seconds per test would take 1^{27} years to complete and yield only coarse insight into parameter interactions. The MiniTab statistical software can design fractional factorial studies, but is typically limited to 15 factors and requires knowledge of each factor's range. Although most of the 71 parameters were defined as C++ floats with the range $[-3.4E^{38}, 3.4E^{38}]$, each parameter's effective range where the prediction works was unknown.

Given the large numbers of parameters in the general aviation prediction, experiments were conducted in a two phased approach. Phase 1 involved an initial set of tests to estimate the range of each parameter, which required the development of a test framework and metrics. Phase 1 tests perturbed parameter values ± 1 order of magnitude from the nominal configuration. Testing the neighborhood of nominal values increased the likelihood of nonzero response metrics from the prediction since zeroed metrics indicated a prediction failure. These tests would inform whether any parameters could be factored out due to negligible effects and provide some indication of how predictions changed, which in turn could be used to validate more detailed results later.

Phase 2 was the production run involving a very large number of tests to estimate each parameter's effective range, effects, and interactions. Due to the large number of tests, the process had to be resumable in the event of a system crash. Phase 2 tests probed each parameter to identify the boundaries where the prediction failed or converged. Since there was no prior knowledge about parameter ranges, each parameter was probed from the nominal value to negative infinity and from nominal to positive infinity, which was repeated for each maneuver and each plugin.

In addition to custom software required to orchestrate automated tests, the iGCAS hazard monitor and prediction plugins required instrumentation. EVAA needed only configuration changes; namely the removal of iGCAS since the orchestration software would call it. iGCAS, in turn, called the configured prediction. The test orchestration software internally instantiated EVAA, set initial conditions, and initialized the mapping system. Within the main loop, the software would perform a parameter update, run a test, and check termination conditions that were intended to avoid wasting time on futile configurations. Metrics were logged to a file using EVAA's built in Logger. The initial metric was a minimum bounding volume described by the tuple ΔX, ΔY, and ΔZ in feet.

EVAA's API for prediction plugins defines a `doPrediction()` function that returns a vector of `OutputModelState`, which is a struct defined in EVAA's API for storing the predicted vehicle position, attitude, speed, and other information for each point in the trajectory. Since that vector is returned to the monitor, capturing metrics in the monitor would preclude having to modify the prediction plugin. However, adding instrumentation to the prediction would be monitor agnostic and make other state information specific to each prediction available that could be exploited for more expressive, albeit prediction-specific, metrics. The latter design was adopted.

The achievability of some trajectory is limited by vehicle dynamics while the extent to which a prediction can be manipulated is limited by its parameters. There are practical limits on how much change a prediction can accommodate. For this reason, control limits set to $\pm 75\%$ of nominal were shown on graphical results. These control limits were arbitrary, but narrower bounds may have required rerunning all tests while broadening the bounds to 100% would have zeroed out all lower parameter bounds.

Prediction parameters are attached to monitors and are loaded by the plugin's public `init()` method, which is automatically called by EVAA's flight-executive on bootup. A simple solution for changing parameters was to simply swap in and reload different configuration files. Various configuration files could be generated using scripting and the `jq` utility that provides a command line interface (CLI) to change specific values in a JSON file.

Swapping configuration files could be implemented quickly and satisfy Phase 1 test requirements, but would not run or scale efficiently. To satisfy Phase 2 requirements, prediction parameters needed to be changed in memory at runtime. This required changing the scope of plugin data structures that characterized different maneuvers and the addition of a public setter for changing specific

Figure 5: GA Left and Straight-Climb Predictions Using ±1 OoM of Nominal Values

parameter values. Phase 2 tests divided each parameter range into a number of equal increments.

The orchestration software used one of four user-selectable parameter update strategies. The first was a binary search that progressively bisected a C++ float's range to find a target value where the prediction would fail or converge, which is similar to the process used in (Reference 14). Next was a low resolution scan that probed the range $\pm 3.4^{38}$ using orders of magnitude from nominal. A medium resolution scan divided a previously determined range into an arbitrary number of equal increments. Finally, a high resolution scan probed the range $\pm 3.4^{38}$ using multiples of the nominal value (or 1 if the nominal value was 0). Each probing operation began by resetting the iGCAS configuration to nominal values to ensure only one parameter was manipulated at a time.

Post processing was comprised of a script that extracted and reformatted metrics from the EVAA log file into a CSV with records containing: configuration code, run number, nominal value, test_X, Y_distance, Y_heading, target_distance, and target_heading. The CSV file would then be run through a Matlab script that determined the effective range for each parameter and also produced a file to generate plots and tables. The effective range was determined by processing a set of records for each parameter-maneuver combination, down-selecting to a contiguous set of records where the prediction response was nonzero and where the response had not converged by ensuring differences between neighboring responses. From that subset, the minimum and maximum values were captured for test_X, Y_distance, and Y_heading.

Results

Phase 1 identified 16 parameters with no effect on the output, 12 of which were not even used in code. The initial results warranted changes to the experimental procedure. Results indicated that most parameters were non-linear and non-monotonic, which meant the binary-search strategy could not be used. The original metrics failed to the capture complex trajectories shown in Figure 5 and were therefore replaced with total-distance-traveled and total-heading-change. Finally, early termination was ultimately disabled to avoid missing any salient features in the more detailed probes.

Excluding 16 factors reduced the dimensionality of the computation problem, but a full factorial set still required 3^{54} tests and there were still too many parameters to generate a robust fractional factorial design using MiniTab. This required re-scoping the experiments. It was only practical to study first order effects for the 49 real valued parameters and second order effects for 5 Boolean parameters. The second order tests toggled one Boolean parameter and repeated a detailed scan. This test strategy afforded the opportunity to include the evaluation of the fixed-airspeed plugin and also assess the effect of varying initial conditions. In all, over 2M iGCAS simulations were executed over a two month period of which results from 1.3M runs were retained and post-processed.

Several hundred pages of tables and plots were produced and, for brevity, were excluded here but can be obtained by emailing the corresponding author. One set of tables list the effective range for each plugin's parameters and represent the prediction's computational limits, i.e., where the prediction yields a nonzero and non-converged response. Parameter effects with respect to each maneuver and each response metric were also tabulated in order of descending effect and sensitivity. On those plots, note that the ΔX and ΔY notation are as defined in SixSigma as a change in input and corresponding change in output. Since most parameter effects are nonlinear, these tables must be used in conjunction with corresponding plots when tuning to ensure perturbed values have continuous responses.

Integrators can exploit these tables using the SixSigma design of experiments workflow, where users select a combination of parameters and their coefficients that yield the target response. Users would have to consult plots that illustrate parameter effects in greater detail to validate and tune the response and ensure effects are not contraindicated. As the response is fine tuned, the figures are again consulted to ensure no unexpected discontinuities or jumps in the response.

The Simplex algorithm[37] transformed these tables into a linear programming model utilizing slack variables, tableaus, and pivot variables. MATLAB user-defined scripts were employed to calculate the linear inequalities for each variable using these tables with respect to each of the four outputs: climb heading, climb distance, roll heading, and roll distance. Given the linear inequalities, a Simplex tableau was implemented in Excel to perform row operations to maximize the linear programming model and check for optimization using OpenSolver. The resulting parameters provided a tuning solution that met the desired setpoints of the four outputs. However, due to time constraints, future work would need to be completed to optimize the Simplex workflow.

CONCLUSION

The EVAA framework has a manageable number of parameters that give integrators flexibility in terms of process allocation, priority, scheduling period, and telemetry rate. The architecture is well-suited for SWAP-C constrained systems and predictable environments and the telemetry stream is suitable for flight-test.

With regard to sensitivity to the underlying computing hardware, we found workload cycle times were principally affected by CPU clock speed but only to some extent. That means faster hardware may not necessarily enable deployment on faster vehicles and that there may be a need for plugins appropriate for different vehicle classes. We also found that the underlying storage or communication technology may use protocols that are CPU intensive and impact workload cycle times. Users intending to minimize cycle times should select processors with simpler instruction sets, single channel RAM, and select peripheral devices or device drivers that minimize their load on the CPU. Users intending to maximize resource utilization should select dual channel RAM and similar underlying technologies designed to increase throughput rather than timeliness.

NASA provided prediction plugins involve a large number of parameters whose behavior can not be verified using exhaustive methods, which makes those components complex systems by definition. Some of the parameters were extremely sensitive, i.e., hyper parameters, that yielded very large changes in the response given small changes in input. However, the complexity of the tuning task for those existing plugins can be simplified by exploiting the structure of maneuver configuration files, which have a section for default parameters and separate sections specific to each maneuver (e.g., Roll, Climb) that override those defaults. A partitioning strategy that isolates parameters specific to maneuvers would reduce the number of parameters under consideration. For example, all bank-related parameters can be moved from the default section to the roll maneuver section to preclude their effect on the climb maneuver. However, this data management strategy does not prevent misconfiguration. The alternative is to implement custom, simpler, predictions.

Because nearly all maneuver parameters were unconstrained and defined as C++ floats, we determined the broadest parameter bounds by perturbing parameter values until response metrics converged or indicated a prediction failure. We narrowed parameter ranges by constraining response metrics to some percentage of nominal and further narrowed ranges to where they were well behaved. The resulting tables and plots from the experiments can be exploited to pick parameter values to tune the predictions to various vehicles.

FUTURE WORK: ARCHITECTURAL IMPROVEMENTS

EVAA's plugin capability decouples the framework from specific vehicles, payloads, and monitors. Its design allows components to efficiently share memory and seeks to minimize SWAP-C requirements, which makes it well-suited for small UAS and simple environments. However, the architectural and IPC features that make EVAA so efficient for SWAP-C constrained systems also limits its deployment to a single computing host and thereby limits the number of plugins and by extension the complexity of the operational environment. As noted in (Reference 10), tuning can not resolve architectural limitations.

For larger Group 3 aircraft hosting a virtualized multi-tenant environment, the challenge is utilization, which may require distributing processes across hosts. Without diminishing the benefits of component reuse, this can be achieved using a micro-service architecture where processes are implemented as stand-alone services that coordinate via messages. This approach would also avoid comingling safety critical and non-safety critical processes since processes can be allocated to hardware with a commensurate assurance level. The micro-service architecture does not necessitate the distribution of processes; all processes can still be allocated to a single computing resource if required. However, note that this architecture does constrain inter-process communication (IPC) options such as shared memory or a shared process address space.

Similar to an operating system, The EVAA-CORE framework does nothing but boot-up to provide an environment for other processes to run. Accordingly, EVAA implements its own form of process management. Transitioning to micro-services would delegate process management to a certified operating system. An additional benefit of this would be the ability to start, stop, restart, or scale processes throughout a mission, whereas EVAA is limited to loading processes only on start-up. This capability would also allow workloads to be modulated by context to improve resource utilization and allow an even greater number of capabilities on board to support complex environments.

Since users can implement plugins with varying levels of engineering rigor, it is absolutely possible for a faulty plugin to crash EVAA in its entirety. A micro-services architecture provides better fault isolation. Each micro-service can also be backed by a supervisor or kubernetes like system for restarting or scaling services, respectively.

This micro-service proposal may address specific limitations as noted here, but to avoid a generate-and-test approach to system architecture there needs to be a study to capture the desired non-functional qualities across different use cases and industries to inform design. This higher level evaluation of architectural capabilities and limits would indicate either a generic architecture or the need for specialized variants. The the costly effort to build a certifiable version of RTA should begin after this architectural question is resolved.

EVAA's built-in telemetry stream is fundamentally limited by the polling design that periodically sends all data to all clients. The substitution of the CVT with a pub/sub data distribution service (e.g., Apache Kafka or RabbitMQ) would not only address the data relevance issue, but allow EVAA's state, sensor, trajectory, and hazard information to be shared with other on-board systems. Replacing EVAA's private database with a public database may also reduce the total number of point-to-point connections between systems, which would improve modifiability and extensibility. Finally, rather than periodic data dumps, a pub/sub system would perform event-driven updates. Note, however, that a pub/sub system may be inappropriate for sensor-coupler plugins for high-data rate sensors since middle-ware and additional network hops will adversely impact performance.

NASA's GA prediction plugin has incredibly high model fidelity taking into account a broad range of factors such as weight, altitude, climb-rate, lateral winds and downdrafts when predicting paths for accelerated maneuvers. However, more work needs to be done on the design of prediction models that support tuning and verification and validation. Over parameterization and unconstrained parameter ranges leads to intractably large and unknowable state spaces.

Finally, high fidelity simulation environments are needed to further explore the performance EVAA or other of autonomy frameworks in highly contentious, adversarial, and multi-agent environments. That capability would support the development and evolution of complex or AI-enabled control, perception, and decision plugins.

ACKNOWLEDGMENT

This six month effort was funded by Modern Technology Solutions Incorporated's IRAD program and managed by Dean Barten, who runs the autonomy lab for MTSI's Warfighter Strategy and Integration group. The authors thank Ethan Williams, the Software Architect at NASA's Armstrong Flight Research Center during the Resilient Autonomy JCTD, for his peer review and comments. The authors also thank Dr. David Elizondo from the De Montfort University Centre for Computational Intelligence (CCI) and Interdisciplinary Group in Intelligent Transport Systems (DIGITS) for his review and feedback.

References

[1] M. T. Jones, *Artificial intelligence: a systems approach.* Laxmi Publications, Ltd., 2008, pg. 369.

[2] D. Seto, B. Krogh, L. Sha, and A. Chutinan, "The simplex architecture for safe online control system upgrades," in *Proceedings of the 1998 American Control Conference. ACC (IEEE Cat. No.98CH36207)*, vol. 6, 1998, pp. 3504–3508 vol.6.

[3] S. Bak, K. Manamcheri, S. Mitra, and M. Caccamo, "Sandboxing controllers for cyber-physical systems," in *2011 IEEE/ACM Second International Conference on Cyber-Physical Systems*, 2011, pp. 3–12.

[4] J. Schierman, M. D. DeVore, N. Richards, N. Gandhi, J. Cooper, K. R. Horneman, S. Stoller, and S. Smolka, "Runtime assurance framework development for highly adaptive flight control systems," 2015.

[5] J. G. Fuller, L. Hook, N. Hutchins, K. N. Maleki, and M. A. Skoog, "Toward run-time assurance in general aviation and unmanned aircraft vehicle autopilots," in *2016 IEEE/AIAA 35th Digital Avionics Systems Conference (DASC)*, 2016, pp. 1–9.

[6] "Standard practice for methods to safely bound flight behavior of unmanned aircraft systems containing complex functions," ASTM International, Tech. Rep., 2017.

[7] L. R. Hook, M. Skoog, M. Garland, W. Ryan, D. Sizoo, and J. VanHoudt, "Initial considerations of a multi-layered run time assurance approach to enable unpiloted aircraft," in *2018 IEEE Aerospace Conference*, 2018, pp. 1–11.

[8] M. A. Skoog, L. R. Hook, and W. Ryan, "Leveraging astm industry standard f3269-17 for providing safe operations of a highly autonomous aircraft," in *2020 IEEE Aerospace Conference*, 2020, pp. 1–7.

[9] H. Koziolek, "Performance evaluation of component-based software systems: A survey," *Performance Evaluation*, vol. 67, no. 8, pp. 634–658, 2010, special Issue on Software and Performance.

[10] Issamjebreen and M. M. Awad, "Preliminary study of software performance models," *International Journal of Advanced Computer Science and Applications*, vol. 7, 2016.

[11] B. Douglass, *Real Time UML: Advances in the UML for Real-time Systems*, ser. Addison-Wesley object technology series.

[12] J. R. Bammi, W. Kruijtzer, L. Lavagno, E. Harcourt, and M. T. Lazarescu, "Software performance estimation strategies in a system-level design tool." New York, NY, USA: Association for Computing Machinery, 2000.

[13] M. Woodside, D. Petriu, J. Merseguer, D. Petriu, and M. Alhaj, "Transformation challenges: from software models to performance models," *Software & Systems Modeling*, vol. 13, pp. 1529–1552, 10 2013.

[14] R. Racu, A. Hamann, and R. Ernst, "A formal approach to multi-dimensional sensitivity analysis of embedded real-time systems," in *18th Euromicro Conference on Real-Time Systems (ECRTS'06)*, 2006, pp. 10 pp.–12.

[15] J. S. Warford, *Computer systems*. Jones & Bartlett Publishers, 2009.

[16] F. Salewski and S. Kowalewski, "The effect of hardware platform selection on safety-critical software in embedded systems: Empirical evaluations," in *2007 International Symposium on Industrial Embedded Systems*, 2007, pp. 78–85.

[17] W. Gu, Z. Kalbarczyk, and R. Iyer, "Error sensitivity of the linux kernel executing on powerpc g4 and pentium 4 processors," 01 2004, pp. 887– 896.

[18] C.-C. Lo, J.-G. Luo, and M.-D. Shieh, "Hardware/software codesign of resource constrained real-time systems," in *2009 Fifth International Conference on Information Assurance and Security*, vol. 1, 2009, pp. 177–180.

[19] J. Roh, K. Lee, J. Lee, S. Oh, H. Kim, and J. Jang, "Development of hil simulator for the sensitivity analysis of factors influencing esc performance," in *2009 ICCAS-SICE*, 2009, pp. 4058–4062.

[20] J. Dominguez-Trujillo, K. Haskins, S. J. Khouzani, C. Leap, S. Tashakkori, Q. Wofford, T. Estrada, P. G. Bridges, and P. M. Widener, "Lightweight measurement and analysis of hpc performance variability," in *2020 IEEE/ACM Performance Modeling, Benchmarking and Simulation of High Performance Computer Systems (PMBS)*, 2020, pp. 50–60.

[21] M. L. George, J. Maxey, D. T. Rowlands, and M. Upton, *Lean six sigma pocket toolbook*. McGraw-Hill Professional Publishing, 2004.

[22] L. Xin, J. K. Becker, S. Gvozdenovic, and D. Starobinski, "Benchmarking the physical layer of wireless cards using software-defined radios." New York, NY, USA: Association for Computing Machinery, 2019.

[23] B. Tierney, E. Kissel, M. Swany, and E. Pouyoul, "Efficient data transfer protocols for big data," in *2012 IEEE 8th International Conference on E-Science*, 2012, pp. 1–9.

[24] F. Ming, J. Li, and H. Zhao, "Design of real-time distributed system using dds," in *International Conference on Automatic Control and Artificial Intelligence (ACAI 2012)*, 2012, pp. 882–885.

[25] N. Cowzer and P. Nixon, "Analyzing implicit group messaging: A novel messaging paradigm for group-oriented content distribution," in *2010 IEEE Globecom Workshops*, 2010, pp. 1518–1522.

[26] A. B. Forouzan, *Data communications & networking (sie)*. Tata McGraw-Hill Education, 2007.

[27] M. Polese, F. Chiariotti, E. Bonetto, F. Rigotto, A. Zanella, and M. Zorzi, "A survey on recent advances in transport layer protocols," *IEEE Communications Surveys & Tutorials*, vol. 21, no. 4, pp. 3584–3608, 2019.

[28] K. Kim, J. Qian, Z. Zhang, Q. Zhou, K.-D. Moon, J.-H. Park, K.-R. Park, and D.-H. Kim, "A scheme for reliable real-time messaging with bounded delays," in *2010 13th IEEE International Symposium on Object/Component/Service-Oriented Real-Time Distributed Computing*, 2010, pp. 18–27.

[29] "Udt: Udp-based data transfer for high-speed wide area networks," *Computer Networks*, vol. 51, no. 7, pp. 1777–1799, 2007, protocols for Fast, Long-Distance Networks.

[30] C. Ionescu, E. Berki, and J. Nummenmaa, "Applying weighted finite state machines to protocol performance analysis," in *2009 Fourth South-East European Workshop on Formal Methods*, 2009, pp. 40–45.

[31] L. Li, C. Zhang, and Y. Ji, "Performance analysis and comparison of mobile instant messaging frameworks," in *2011 7th International Conference on Wireless Communications, Networking and Mobile Computing*, 2011, pp. 1–4.

[32] B. Ciuffo and C. Lima Azevedo, "A sensitivity-analysis-based approach for the calibration of traffic simulation models," *IEEE Transactions on Intelligent Transportation Systems*, vol. 15, no. 3, pp. 1298–1309, 2014.

[33] Q. Ge, B. Ciuffo, and M. Menendez, "An exploratory study of two efficient approaches for the sensitivity analysis of computationally expensive traffic simulation models," *IEEE Transactions on Intelligent Transportation Systems*, vol. 15, no. 3, pp. 1288–1297, 2014.

[34] V. Punzo, M. Montanino, and B. Ciuffo, "Do we really need to calibrate all the parameters? variance-based sensitivity analysis to simplify microscopic traffic flow models," *IEEE Transactions on Intelligent Transportation Systems*, vol. 16, no. 1, pp. 184–193, 2015.

[35] M. G. Mehrabi, R. M. H. Cheng, and A. Hemam, "Parametric study and sensitivity analysis of automated vehicles," *Robotica*, vol. 13, no. 5, p. 469–475, 1995.

[36] R. V.S.S. and V. R., "Sensitivity analysis of the turning motion of an underwater glider on the viscous hydrodynamic coefficients," *Defence Science Journal*, vol. 71, no. 5, p. 709 – 717, 2021.

[37] S. I. Gass, *Decision making, models and algorithms.* John Wiley & Sons, Inc., One Wiley Drive, Somerset, NJ 08873 (USA), 1985.

Inertial Measurement Unit for Robust Control and Guidance of UAVs

Tehmoor Dar[1], Ph.D
Thales Defense & Security, Inc.

An Inertial Measurement Unit (IMU) is a device that uses accelerometers and gyroscopes to measure the acceleration and rotation of an object to which it is attached. Some IMUs also include a magnetometer to measure magnetic strength and a central processing unit that embeds a state observer to calculate attitude and heading relative to magnetic north. The output from an IMU aided with a GPS/GNSS receiver (integrating dead reckoning) is commonly used to calculate orientation data, roll pitch, and yaw faster with higher accuracy. Inertial Navigation System (INS) is widely used for motion tracking and monitoring.

Precision, accuracy, and robustness are critical parameters that drive the selection of an IMU for challenging applications, ranging from deep-sea drilling to Unmanned Aerial Systems (UAS). These applications require a robust and stable IMU that can compensate for extreme temperature ranges and a highly dynamic vibrational environment. A robust output from an IMU will help the UAS with easy control and guidance.

There are different types of IMUs available in the market based on technologies such as Micro-Electro-Mechanical Systems (MEMS), Fiber Optic Gyroscopes (FOG), and Ring Laser Gyroscopes (RLG). MEMS technology is the most cost-effective in terms of size, weight, and power (SWaP) when comparing these technologies for applications such as Unmanned Aerial Vehicles (UAVs).

This presentation will focus on how to evaluate an IMU in terms of technologies, robustness, and performance for UAVs. In addition, it will also share techniques to mitigate the impact of extreme conditions on the performance of IMUs using various statistical estimation algorithms and hardware implementation.

INTRODUCTION

Regardless of their size and mission, all UAVs need sensors to estimate their complete trajectory. The trajectories or state vectors in the estimation model are explained mathematically by critical parameters such as position coordinates, components of the velocity vector, and parameters that describe the vehicle's attitude. The control and guidance system uses these critical parameters to maneuver the UAV as per its mission. The guidance system generates instructions for the UAV to follow its predestined path, and the navigation block determines the vehicle's location, velocity, and attitude. At the same time, the control

[1] Principal Engineer, Thales Defense & Security, Inc., 700 Technology Park Dr. #102, Billerica, MA 01821

system manipulates the forces by steering controls, thrusters, etc., needed to track guidance commands while maintaining vehicle stability. A high-level block diagram capturing the flow of a Guidance, Navigation, and Control (GNC) system is shown in figure 1.

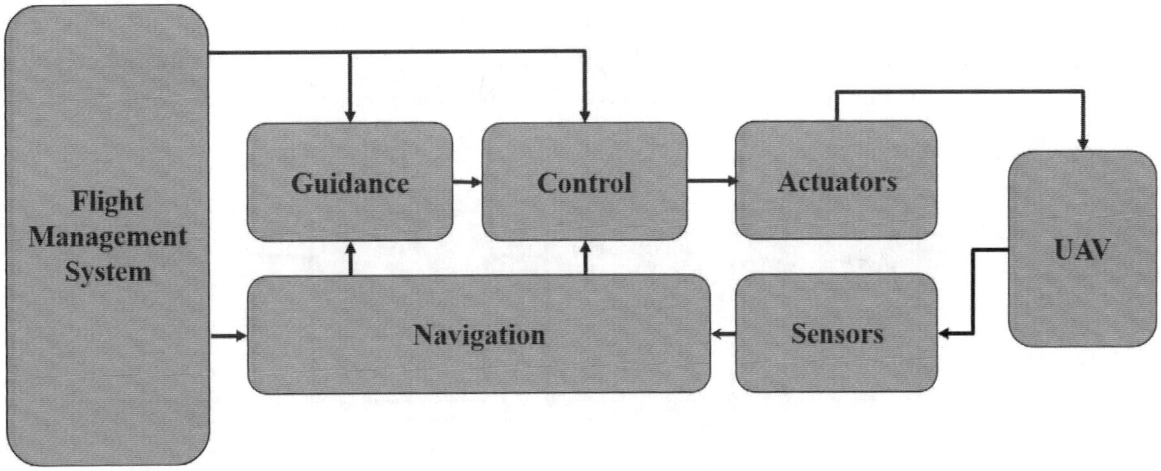

Figure 1 GNC Block Diagram

GNC implements the state estimation in the UAV. The controllers require very high-resolution and precise data to control the UAV, for example for an accurate landing. Sensor fusion is a methodology using onboard computers to achieve GNC. It overcomes the limitations of individual sensors by gathering and fusing data from multiple sensors to produce more reliable data with less uncertainty. In UAVs, any electronic component's size, weight, and power are critical. Therefore, it is vital to achieve sensor fusion data with a minimum number of sensors.

Some of the most critical parameters to define the complete trajectory of a UAV are position, velocity, attitude, and rotation (pitch, roll, and yaw) rates. Position, velocity, and attitude are also known as the navigation state. The sensors used to measure these quantities are called navigation sensors. The two most common navigation sensors are an inertial measurement unit (IMU) and a global navigation satellite system (GNSS). An inertial navigation system (INS) computes the states using only data from an IMU. An IMU for robust control and guidance of UAVs is the focus of this paper.

As discussed, an INS uses a computer, motion sensors (accelerometers), and rotation sensors (gyroscopes) to continuously calculate by dead reckoning the position, the orientation, and the velocity of a moving object without the need for external references. A complete six-degree-of-freedom inertial sensor consists of 3-axis accelerometers and 3-axis gyros. The accelerometer measures the specific force acting on the platform, and the gyro measures its rotation. However, these sensors are susceptible to errors because of various manufacturing processes. These position, velocity, and attitude estimates errors are mainly due to sensor bias and noise corrupting the measurements.

Inertial Measurement Unit

The success of a UAV depends on accurate position and motion estimation through challenging environments. IMU data is usually fused with GNSS beside vision and other ranging systems to navigate known and unknown terrain. However, UAVs must accurately estimate where they are and where they are heading, even when the GNSS signal is unavailable.

As discussed in the introduction, the IMU is the most critical sensor to guide and control the UAV trajectory. The IMU measures the acceleration and rotational velocity of the UAV in action. An IMU is the most dependable set of sensors as it can neither be jammed nor affected by any environmental condition. The IMU data is always available and integral to any UAV GNC algorithm.

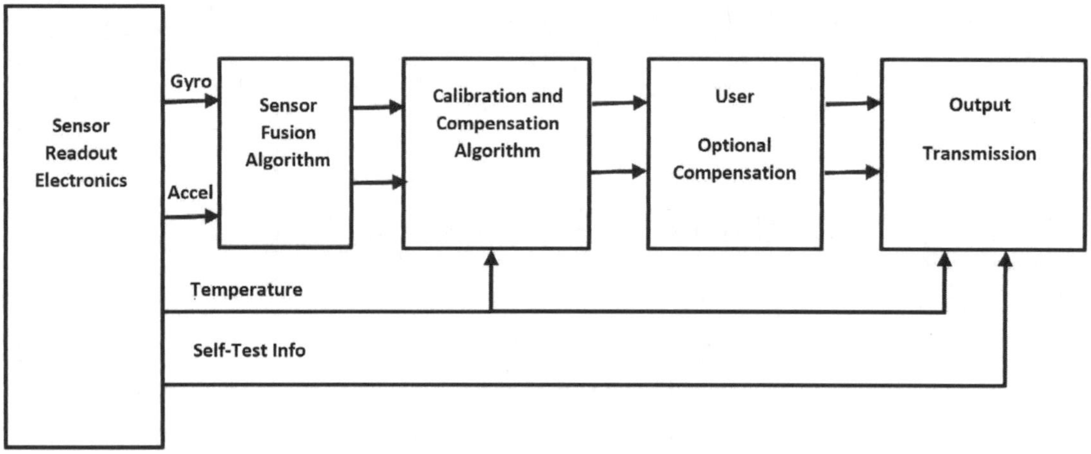

Figure 2 IMU Block Diagram

As shown in figure 2, a typical IMU block diagram includes a gyroscope, an accelerometer, a temperature sensor, and supporting algorithms. An IMU typically includes A/D conversion to extract the measurements, temperature compensation, and filtering to improve the output. Sensor fusion numerically integrates and solves the data for rotation angle ($\Delta\Theta$) and linear velocity (ΔV). Calibration and compensation correct any misalignment or sensor biases. A user has an option to rotate ($d\Theta$) and (dV) from IMU module's internal axis to the mounted body (Vehicle's) frame of reference before transmitting the final data.

Inertial Measurement Unit Technologies

There are many types of IMU, some of which incorporate magnetometers to measure magnetic field strength, but the four main technological categories for UAV applications are:

- Silicon MEMS (Micro-Electro-Mechanical Systems),
- Quartz MEMS,
- FOG (Fiber Optic Gyro), and
- RLG (Ring Laser Gyro).

Silicon MEMS IMUs, because of their size, weight, power, and lowest manufacturing cost are ideal for small UAVs. Silicon MEMS are designed to measure the deflection of a mass during any platform motion. They intrinsically have higher noise output and are sensitive to vibration, and they also have higher instability parameters than more expensive FOG IMUs. But, recently, we have seen improvement in these critical parameters as technology continues to develop.

The FOG IMU is a solid-state technology design. It uses optical fiber loops and measures interference in beams of light in opposite loops to detect rotation in each axis. It is less sensitive to shock and vibration and offers excellent thermal stability. It is susceptible to magnetic interference and consumes more power. It outperforms MEMS IMUs in critical parameters such as angle random walk, bias instability, and bias offset error, making it ideal for mission-critical UAV applications.

An RLG IMU utilizes solid-state technology and operates like a FOG with a sealed ring cavity. It is the most stable and accurate IMU, but it is also the biggest and most expensive compared to other IMU technologies.

Quartz MEMS IMUs use a one-piece inertial sensing element, micro-machined from quartz, driven by an oscillator to vibrate at a precise amplitude. The vibrating quartz is used to sense angular rate. These are very comparable in performance to FOG and RLG IMUs and thus ideal for larger size UAVs.

IMU Performance Grade

IMUs for UAVs need to be highly rugged to withstand high levels of shock and vibration and wide temperature ranges. IMUs with low vibration sensitivity provide higher accuracy navigation solutions in challenging UAV applications where vibration levels are high.

Inertial sensors are generally categorized into performance classes according to bias stability specifications. The higher the sensor grade, the more accurate it tends to be, as shown in figure 3. Typically, tactical- and industrial-grade IMUs are preferred for challenging applications such as UAV navigation, as they offer high performance and accuracy. Tactical-grade IMUs have gyroscopes with low in-run bias stability, an offset error that is more stable over time. The industrial-grade IMUs provide lower accuracy than tactical-grade-IMUs, but they offer good tolerance to shock and vibration and exhibit good repeatability over time and temperature. MEMS-based IMUs are commonly used for applications requiring small form factors such as low-cost UAV applications.

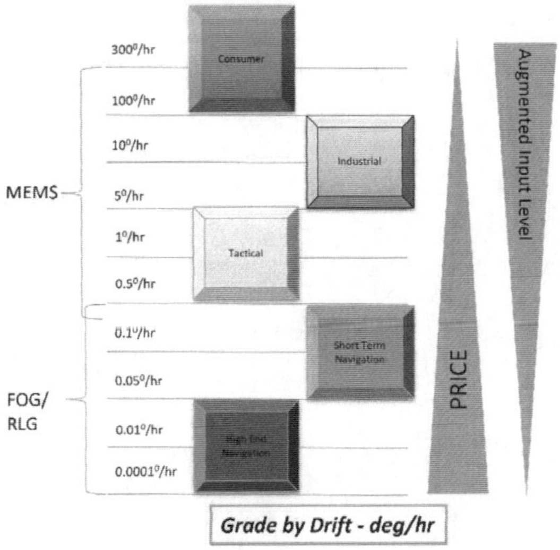

Figure 3 Sensitivity Performance Grade IMU

IMU Selection Strategy

Engineers designing systems for UAV applications need to define and prioritize the performance, SWaP, and cost factors that are most important to the success of their design to select the appropriate IMU and technology. Comparing IMUs based on their specific technologies and performance strengths and weaknesses will be vital for the design process.
The performance of an IMU is dependent on the sensor technology and the thermal properties of the packaging. There are numerous parameters based on applications that define the accuracy and stable performance of the IMU.

The following parameters are the most critical when comparing the performance and accuracy of specific IMUs, to help determine suitability for a given UAV application:

- Bias – The output of an IMU when input is zero.
- Bias repeatability – The change in IMU bias measurement when conditions have changed, such as for each power-up of the IMU.
- Bias stability - The measurement of the change in bias over time.
- Random Walk – The random noise present in the output.
- Vibration Sensitivity – The change in the inertial output per vibration unit present in the environment.

Allan Variance Test for Performance Parameters

The Allan variance is a method to estimate frequency stability in an oscillator due to noise. This method is used for identifying and quantifying the different noise terms that exist in inertial sensor data as a function of the averaging time. These noise terms in the inertial sensor data appear in the log-log plot of $\sigma(\tau)$ with the different slopes as shown in figure 4. To create such a plot, a sensor is mounted statically inside a thermal chamber set to a constant temperature, and data is logged at a high rate for an extended period. The data is down-sampled at various time constants, and the variance at each time constant is calculated and plotted. Despite its name, most of the time, the results are plotted as standard deviations (square root of variance) versus the down-sampled time constants on a log-log scale.

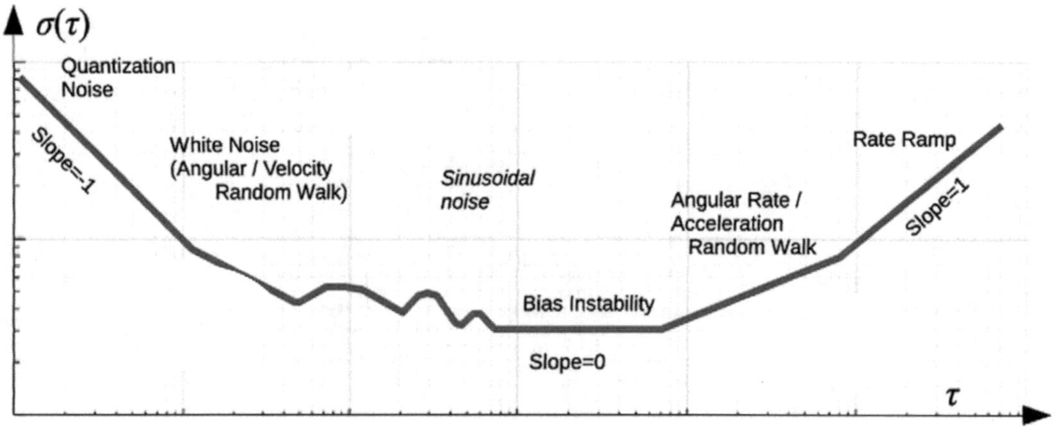

Figure 4 Allan Variance Chart

The Allan Deviation plot in figure 4 illustrates some critical specifications for an inertial sensors. Autonomous system designers work with two crucial parameters to improve MEMS gyroscopes: angle-random-walk (ARW) and bias instability (BI). These parameters provide designers with a standard benchmark to compare the relative performance of gyroscopes and determine the time envelope of safe operation:

- Angle random walk is a measure of short-term noise in the rate sensor. It is commonly expressed in degrees per root hour and is calculated from the one-second point (Tau = 1sec) on the Allan variance (deviation) plot of an inertial sensor output. It is essential information to an INS because an algorithm integrates the inertial data to determine the vehicle's attitude (roll, pitch, yaw), and integration of noise results in a random error.
- Bias instability (BI) is a measure of long-term noise in the rate sensor. It is also derived from the Allan variance plot of an inertial sensor output. Commonly expressed in degrees per hour, it is calculated from the minimum point on the Alan Variance curve and represents the long-term drift of the gyro bias. During the integration of the rate signal, any error in the bias will quickly accumulate as angle error.

Gyroscopes Acceleration Sensitivity

A gyroscope would only measure the angular rate and have no linear acceleration sensitivity in an ideal world. In practice, gyroscopes are susceptible to linear accelerations due to asymmetrical design and manufacturing tolerances. Two types of linear acceleration sensitivities impact UAV performance critically, and these are

- G-Sensitivity is an acceleration sensitivity in which a gyroscope experiences a bias shift when subjected to a constant linear acceleration. A gyroscope's bias sensitivity to linear acceleration can be calibrated in real-time using well-calibrated accelerometer values.g^2 sensitivity is an acceleration sensitivity that causes a bias shift in the output of a gyroscope due to oscillatory linear accelerations. Gyroscopes' sensitivity to acceleration can be readily determined through vibration testing, using sine-sweeps or random vibration profiles.

The most common method to compensate for g-sensitivity is using an external accelerometer like IMU applications, and another common strategy to improve g2-sensitivity is to add a mechanical anti-vibration mount to attenuate narrow-band vibration.

Sensor Fusion Algorithm Significance

Advanced sensor fusion using state estimation techniques like Kalman filtering can correct ARW and BI errors during regular UAVs operation. Pitch and roll gyroscope errors can be removed within an IMU using the Kalman filter and calculations of pitch and roll made using accelerometer signals measuring acceleration due to Earth's gravity. The algorithm tracks and corrects the bias drift and corrects the ARW error. The impact of accelerometer and gyro bias on position error (vertical axis) due to gravity misalignment is shown in the figure 5.

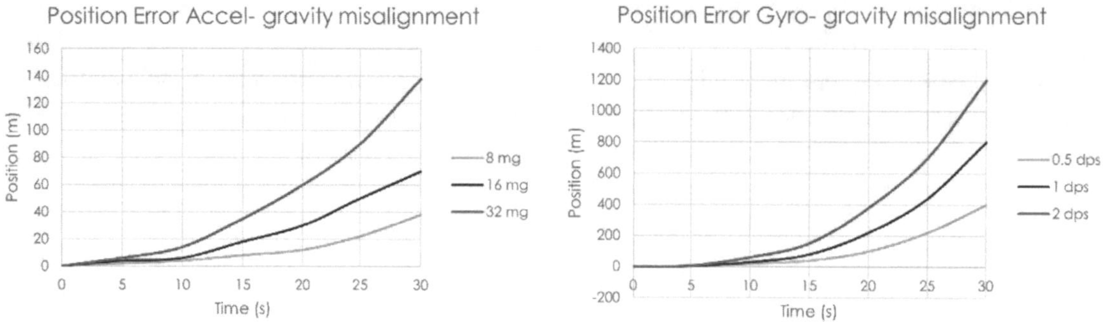

Figure 5 Position Error Accelerometer and Gyroscope

As discussed above, the yaw axis is an essential but different measurement compared to roll and pitch. The yaw axis is along the gravity vector; acceleration signals from the earth's gravity play no role in any heading correction. Instead, reference signals derived from GNSS and other sensors on the UAVs become critical for the heading correction.

In the presence of reference signals from GNSS and other sensors on a UAV, the sensor fusion algorithms track and correct the yaw gyro errors. If the reference signals are lost or compromised, the autonomous system must rely on the raw performance of the yaw gyro signal. In this case, ARW and BI determine how much position, velocity, and heading error will build with time.

Yaw gyro performance remains a constant issue for users. It drives manufacturers to improve gyro sensitivity and overall performance for UAVs with bias instability of significantly less than 1 deg/hr. Highly stable and more accurate IMUs, especially the gyros, will enhance UAVs applications.

Conclusion

We learned that it is critical for UAVs operating without GNSS to select high precision IMUs with a design to reject shocks and vibration with minimum bias instability and noise parameters for robust and repeatable operations. It is essential to design the application with its IMU in the middle of the application board with special design mountings to reject any vibration or shock creating a bias shift. Besides IMU technology selection for UAVs, the design of an estimation algorithm to measure sudden angular rate error and bias shift is also a key feature for any UAVs operating without GNSS.

NMAC DATABASE FOR MACHINE VISION DAA DEVELOPMENT AND ASSESSMENT

Iryna Borshchova[1] and Kristopher Ellis[2]

Many research organizations, companies, and universities are studying the Detect and Avoid problem and attempting to use electro-optical technology to detect and track non-cooperative aircraft. Often these stakeholders resort to the use of synthetic imagery or conduct limited flight tests with intercepts that would not result in a NMAC (Near Mid-Air Collision). While these techniques provide valuable insight into the Detect and Avoid problem, they do not replicate the challenges associated with the detection of a stable NMAC trajectory from a moving aircraft under real conditions. When the intruder and ownship are on a non-accelerating collision path, the intruder remains fixed in azimuth and elevation relative to the camera, which may make it difficult to be "recognized" by the detection/tracking software. Therefore, it is important that Detect and Avoid algorithms are verified on flight test image data collected during real near mid-air collision intercepts.

The National Research Council's (NRC's) Aerospace Research Center has been conducting NMAC intercepts to investigate techniques and technologies to address the unmanned aircraft Detect and Avoid problem since 2008. As part of these investigations, the NRC possesses a library of video imagery collected from 2016 to 2019 with a Bell 205 helicopter acting as a surrogate unmanned aircraft and flying intercepts against multiple fixed-wing intruders (Harvard/Texan, Extra 300, DeHaviland Twin Otter). The imagery is from seven calibrated machine vision cameras mounted on the surrogate unmanned aircraft (ownship), as well as the truth data regarding position information as recorded by the high-quality inertial measurement systems on-board both the ownship and the intruder aircraft.

This paper describes the NRC's NMAC imagery database, including: the camera specifications, extrinsic/intrinsic camera parameters, and camera installations on the surrogate unmanned aircraft. It proposes methods for establishing performance metrics of an electro-optical Detect and Avoid system, which are intended to support emerging standards. The NRC's NMAC database and its associated truth data may become a standard to benchmark electro-optical Detect and Avoid algorithms and is available freely upon request (contact NRC.NMACDatabase-basededonneesNMAC.CNRC@nrc-cnrc.gc.ca for access).

[1] Research Officer, Aerospace Research Centre, Flight Research Laboratory, 1200 Montreal Road, Ottawa, ON, Canada K1A 0R6.

[2] Senior Research Officer, Aerospace Research Centre, Flight Research Laboratory, 1200 Montreal Road, Ottawa, ON, Canada K1A 0R6.

INTRODUCTION

Many research organizations, companies, and universities are studying the Detect and Avoid (DAA) problem and attempting to use electro-optical technology to detect and track non-cooperative aircraft. Often these stakeholders resort to the use of synthetic imagery or conduct limited flight tests with intercepts that would not result in a NMAC (Near Mid-Air Collision).

Reference 1 describes a vision-based DAA system designed to operate on small and medium sized RPASs (Remotely Piloted Aircraft Systems). The intruder detection was done on a 4650×1280 sized synthetic video flow which was processed by a cellular processor array. Reference 2 implemented an approach to detect another RPAS using an onboard monocular camera. The head-on and cross-over collision intercepts were conducted between two RPASs, however, the separation distance between them at the starting point was only 20 m which was far too close a range for UAV operations in commercial airspace. Reference 3 suggested the use of synthetic imagery from desktop simulators such as FlightGear (Reference 4) or X-Plane (Reference 5) for the development and testing of vision-based DAA technologies. However, off the shelf simulation systems do not have the visual fidelity required to simulate a computer vision system accurately. This limitation may be overcome by building visual fidelity simulators using high performance rendering engines (e.g. Unreal) but there is a substantial cost that can put this option out of reach for small and medium enterprises (SMEs). At the same time, SMEs developing vision-based DAA systems often do not have resources and experience in flight test and the safe conduct of actual collision intercepts. In most cases, having two aircraft arrive at a predetermined position at a chosen time, while maintaining a constant geometry, requires the solving of two simultaneous four-dimensional (longitude, latitude, altitude, and time) navigation problems (References 6 and 15). Such encounter trajectories must be simultaneously followed by both the ownship and the intruder aircraft prior to the initiation of the collision avoidance maneuver. Given these challenges, typical DAA flight testing performed by SMEs involves a limited number of collision intercepts and lower fidelity collision intercepts where the intruder aircraft is not on a stable collision path.

While these techniques provide valuable insight into the Detect and Avoid problem, they do not replicate the challenges associated with the detection of a stable NMAC trajectory from a moving aircraft under real conditions. Firstly, intensity and colors of objects on synthetic images differ from the real ones. Secondly, when the intruder and ownship are on a collision path, the intruder remains fixed in azimuth and elevation relative to the camera, which may make it difficult to be detected by the detection/tracking software from far away. Therefore, it is important that Detect and Avoid algorithms are verified on flight test image data collected during real near mid-air collision intercepts.

The National Research Council's (NRC's) Aerospace Research Centre has been conducting NMAC intercepts to investigate techniques and technologies to address the unmanned aircraft Detect and Avoid problem since 2008. As part of these investigations, the NRC possesses a library of video imagery from seven calibrated machine vision cameras mounted on the surrogate RPAS, as well as the truth data regarding position information as recorded by the high-quality inertial measurement systems on-board both the ownship and the intruder aircraft. This data was collected from 2016-2019 with a Bell 205 helicopter acting as a surrogate unmanned aircraft and flying intercepts against multiple fixed-wing intruders (i.e. a Harvard/Texan, an Extra 300 and a DeHavilland Twin Otter).

This paper describes the NRC's NMAC database and is organized as follows: Section 1 provides a detailed description of the database. Section 2 describes the camera array used to collect image data and camera calibration procedures. Section 3 describes the test aircraft and the experiments conducted. Section 4 demonstrates preliminary analysis of the performance metrics of the NRC's image-based detection and tracking algorithm. At the end of the paper, conclusions and recommendations for future work are presented.

1 DATABASE DESCRIPTION

The NRC's NMAC database (Reference 7) is available freely upon request (contact NRC.NMACDatabase-basededonneesNMAC.CNRC@nrc-cnrc.gc.ca for access). The database contains data from eight sets of collision intercepts with the NRC's Bell 205 acting as a surrogate RPAS and three different intruder aircraft (Harvard/Texan, Extra 300, DeHaviland Twin Otter). Specifically, the database includes following folders:

1. A folder with camera calibration parameters for each camera (details on camera sensors are described in Section 2);
2. A folder with images from the seven-camera array for each collision intercept;
3. A folder with "ground truth" data for each collision intercept (details are given in Section 3.2);
4. A folder with "pixel truth" data for each collision intercept (details on data collection are given in Section 3.2).

1.1 Calibration Folder:

A folder with camera calibration parameters consists of seven *.csv* files from each camera and a *.txt* file with a homography matrix. The *.csv* files contain:

1. Camera name (Master/Slave# etc.);
2. Camera port;
3. Width/height of the image in pixels;
4. Focal lengths of the camera in pixels;
5. Coordinates of the optical centre of the image in pixels;
6. Rotational (roll, pitch, heading) and translational (x, y, z) shifts of the camera relative to the Master camera in degrees (details on the installation and calibration are described in Section 2).

1.2 Image Data Folder:

The folder with image data contains images from each camera for each conducted intercept. The data was collected as uncompressed 2448x2050 pixel 8-bit monochrome bitmaps (*.bmp*) with the file-naming convention of a *frame_number.bmp*. The frame counter was restarted at the beginning of each encounter set.

1.3 Ground Truth Folder:

The folder with the ground truth data contains: 1) camera data in the *.srt* format, a text based subtitle format allowing for text data to be overlaid over video images; and 2) *.mat* file with ownship/intruder position data. The Matlab structure contains following fields:

1. Registration of the aircraft;
2. Aircraft type;
3. GPS time (sec);
4. Latitude (degrees);
5. Longitude (degrees);
6. Altitude (ft) and (m);
7. UTM coordinates E/N (m)and zone;
8. Velocities (E, N, Up) – m/s;
9. Ground Speed (knots);

10. Track angle (degrees);
11. Flight path angle (degrees);
12. Roll, pitch, heading (degrees).

The information on camera state and image time for each camera in the *.srt* file includes:

1. Time at which image was taken (Local INS time, Aircraft time, and Camera time);
2. Frame number;
3. Time elapsed since previous image was taken;
4. Camera state (roll, pitch, heading) in degrees;
5. Camera position (latitude, longitude, altitude);
6. Camera velocities in aircraft body frame (rad/sec);
7. Camera accelerations in aircraft body frame (m/sec^2).

1.4 Pixel Truth Folder

The "pixel truth" folder contains:

1. Horizontal pixel truth;
2. Vertical pixel truth;
3. Projected horizontal location of the intruder on the image in pixels;
4. Projected vertical location of the intruder on the image in pixels.

Further details on "pixel truth" data and how it was established is given in Section 3.2.

2 CAMERA ARRAY OVERVIEW

2.1 Hardware architecture

This Section describes PICAS (Passive Intelligent Collision Avoidance Sensor, Reference 8), the NRC's prototype of a camera-based collision avoidance system used for image data collection during collision intercepts. PICAS consisted of passive electro-optical (EO) elements operating in the visible to near-infra-red spectrum. The instrument served as a platform for the development, refinement and validation of vision-based techniques dedicated to detection and tracking of potential collision-course targets.

The PICAS system utilized a multi-camera array comprised of seven nodes (Figure 1 a) and b)) mated to a rackmount PC for data recording and processing (Figure 1, c)). Each node consisted of a Prosilica GC2450 machine vision camera (Reference 9), housing a 2448 x 2050 pixel (5 megapixel) monochrome focal plane array operating at a maximum acquisition rate of 15 Hz.

The optical configuration changed between nodes, resulting in selective sampling across the overall field-of-view (FOV). Each camera was labelled with a number from 0 through to 6, with Camera 0 representing the forward facing camera, and Camera 6 nominally pointing at a -88 degree azimuth. The calibration of each camera node was performed relative to the forward facing Camera 0. Frame synchronization was available as a software configured option with Camera 0 being the frame trigger source. To highlight this fact, Camera 0 was referred to as the Master Camera, and Cameras 1-6 were referred to Slaves 1-6. For the data set described herein, however, all cameras were configured to free-run based on their own internal clock.

The camera array was configured as shown in Table 1. Typical azimuthal overlap was constrained to 15% or less, as minimal overlap resulted in a maximized FOV. However, some overlap was necessary to account for manufacturing tolerances and ease of calibration. The total effective azimuthal coverage of the array, including overlaps, was computed to 110 deg., with coverage in elevation ranging from 11 deg. in the forward-look direction to 34 deg. at the 80-degree azimuth and beyond. A minimal overlap of 2 degrees was provided between nodes to aid in calibration and minimize the possibility of azimuthal gaps arising from the machining and assembly process. *Note* that the azimuthal coverage is a subset of the 360 degrees and represented coverage on one side of the aircraft. As a prototype demonstrator, the PICAS array was adequately configured to test co-altitude collision course geometries along a variety of azimuths.

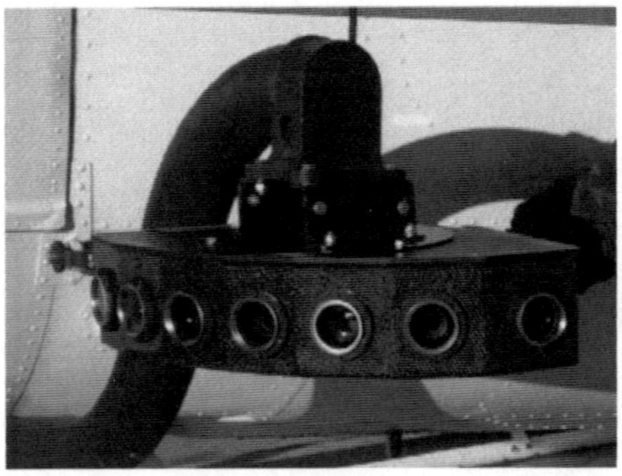

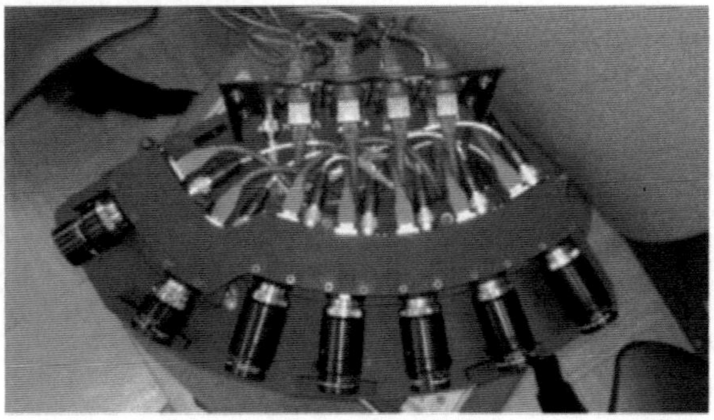

(a) PICAS camera array mounted on the Bell 205 (b) PICAS camera array with seven cameras exposed.

c) PICAS processing computer

Figure 1. PICAS system

Table 1. PICAS Node specification

Parameter	**Unit**	**Master, Slave 1-4**	**Slave 5**	**Slave 6**
Pixel count	pixel	2448 x 2050	2448 x 2050	2448 x 2050
Focal length	mm	35	23	12
FOV	degrees	60.8 x 11.6	21.0 x 17.6	40.3 x 33.8

The field of view for all cameras can be visualized projecting a set of grid points from the image bounds into a common camera reference frame and scaling the points to fit on a unit sphere as seen in Figure 2a. Alternatively, the field of view can be projected onto the image plane as seen in Figure 2b, which is achieved by projecting into image coordinates and normalizing by dividing by the z-component.

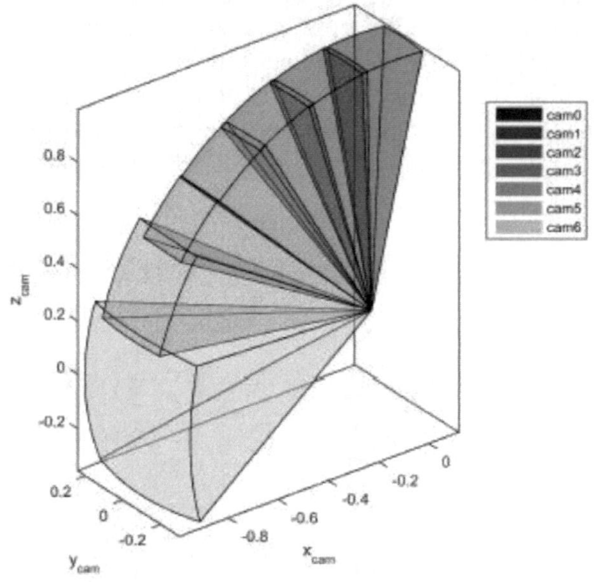

a) Simulated FOV of PICAS on unit sphere relative to Camera 0

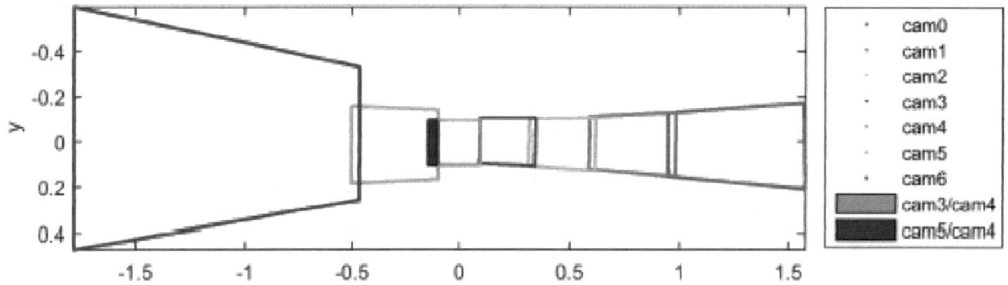

a) Overlapping FOV of PICAS nodes relative to Camera 5

Figure 2. PICAS FOV

2.2 Intrinsic/Extrinsic calibration

Collision-course targets detected by PICAS had to be conveyed to the avoidance system or a pilot warning display in a common coordinate reference frame. Geodetic coordinates were a suitable choice, provided that it was possible to transform the sensor data into a world coordinate reference frame. A calibration scheme was needed to extract that transformation accurately, as errors in the calibration process contribute to the overall error in target position and velocity.

The calibration of an outward-looking multi-camera array such as PICAS is non-trivial. The wide azimuthal field of view (110 deg., Figure 2a), minimal overlap between sensors (Figure 2b), differing fields of view, and lenses focused to infinity made it challenging to apply standard camera calibration techniques. As such, a multi-step camera calibration process that achieves adequate accuracy for a DAA solution was developed and implemented. This approach consisted of intrinsic and extrinsic calibration developed by the NRC under the leadership of John White, details of which are described below.

Intrinsic calibration determined the transform from camera coordinates to the image plane and utilized the calibration toolbox developed by Jean-Yves Bouguet (Reference 10). Extrinsic calibration determined the homographies between the cameras coordinates for every adjacent pair. The exterior transform between any two camera coordinates was calculated by matrix multiplication of the adjacent homographies. The conversion from image coordinates to geodetic coordinates for a camera system attached to a mobile platform necessitated the decomposition of a single transformation into multiple intermediate transformations, as shown in Figure 3 and described in detail in Reference 11.

To establish the intruder's location in the geodetic coordinates the image frame coordinates must had been transformed into geodetic coordinates by first rotating around camera axis (heading, pitch and roll of the camera), then around the body axis (heading, pitch and roll of the aircraft), while accounting for camera intrinsic parameters. This process is described in Figure 3, and is given by the Eq. (1):

$$[Xg,Yg,Zg] = M_{g\leftarrow b} \times M_{b\leftarrow c} \times M_{c\leftarrow I} \times K^{-1} \times [u,v,1]; \quad (1)$$

Where $[u,v,1]$ – pixel coordinates of intruder on the image frame, Xg,Yg,Zg – intruder's coordinates relative to ownship in the Earth fixed frame, $M_{g\leftarrow b}$, $M_{b\leftarrow c}$ – rotational matrices around body and camera axis respectively, $M_{c\leftarrow I}$ -conversion to image frame, K^{-1} – inverse of the interior matrix. This transformation neglects the effects of the lever arm translation between the aircraft IMU reference point and each camera's optical reference point as these factors are negligible compared to other sources of potential error, such as rotational offsets.

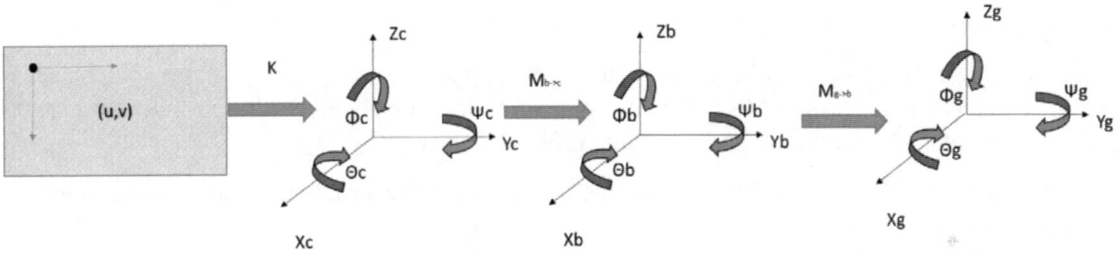

Figure 3. Transformation from pixel to Geodetic

Considering that PICAS is a camera array, each transformation was done in a camera's local frame, such that:

$$M_{b\leftarrow c} = M_{b\leftarrow c_m} \times M_{c_m \leftarrow c_n}; \quad (2)$$

Where $M_{b\leftarrow c_m}$ is the rotation around "Master" camera (Camera 0) a, and $M_{c_m\leftarrow c_n}$ is the transformation from the local camera to the "Master" camera (Reference 12).

The calibration was performed by using a planar target with a checkerboard pattern printed on one face. The target was designed such that it covered the majority of any field of view considered and the images taken of the target provided accurate corner extraction. The dimensions were set by the field of view and scaled by the minimal distance between the target and camera that gave a focused image. Additionally, the squares were small to maximize the number of calibration points per image and were large enough for the corner extraction algorithm to work.

The calibration procedure for PICAS may be divided into four steps. First, images of the calibration target were taken by the PICAS camera array. Secondly, the calibration points were extracted from the images by using the corner extraction tool provided in Bougeut's toolbox (Reference 10). Thirdly, the intrinsic calibration was performed via the calibration tool. Finally, the extrinsic calibration was performed on adjacent cameras by taking advantage of the implicit homography of the planar target.

After the images were collected and corners were extracted, the intrinsic calibration on a set of images from a single camera was performed by running the calibration tool from Bouqeut's toolbox once the corners have been extracted. The calibration toolbox presented initial guess and optimized values with uncertainties for the focal length, principle point, and distortion coefficients. The calibration also returned the extrinsic rotation matrix and translation for each image and built the intrinsic matrix.

Several checks were done to determine the quality of the intrinsic calibration. The focal length was within error of the manufacturer specifications for the lens, and there were no significant variations between the x and y component. Camera 0, 1, 2, 3, and 4 had a lens with a focal length of 35.2 mm, whereas Camera 5 and 6 had lenses with a focal length of 22.5 mm and 12.7 mm respectively. All cameras used the Prosilica GC2450 imager which had an image resolution of 2448 x 2050 pixels, hence the theoretical value of the principle point was $[1224\ 1025]^T$. Deviation from the theoretical value implied that a given perspective had a bias.

The extrinsic calibration determined the exterior transform M (Eq. 1 and 2) that mapped from the coordinate system of one camera to another. Utilizing the planar target, the exterior transform was simplified to a homography that maps the optic plane of one camera to that of another camera. Only the homography between adjacent cameras was calculated directly as the target must occupy the majority of the combined field of view of both cameras. The exterior transform between any two cameras was calculated by multiplying the adjacent transforms and their inverses in a logical sequence. Unfortunately the error associated with each adjacent homography used to build the exterior transform contributed to the transform's uncertainty.

The homography was calculated by using *ransac_thomography vgg.m* from Zisserman's multiple view geometry toolbox (Reference 13). This function utilized additional functions from Kovesi's computer vision and image processing toolbox.

Table 2 presents the results of the inter-camera calibration process, where the nominal (design) heading offset per node, defined by the optic axis, is contrasted against the measured offset from the calibration process. The worst-case calibration error was +/-0.25 deg., driven by the composite homography between the Master and Slave 6, while uncertainty grew on average as we move away from the Camera 0. The average calibration error was within +/-0.1 degrees.

Table 2. Camera heading after calibration

Camera	Nominal heading (deg.)	Calibrated heading (deg.)	Difference between nominal and Calibrated heading (deg.)
Master	0	0	0
Slave 1	12	12.39	0.39
Slave 2	24	25.02	1.02
Slave 3	36	37.30	1.30
Slave 4	48	50.74	2.74
Slave 5	63	65.74	2.74
Slave 6	88	88.47	0.47

3 EXPERIMENTAL RESULTS

3.1 Test aircraft

Flight tests were conducted in uncontrolled airspace near the NRC's facility at the Ottawa International Airport at an altitude of 2000-3000 ft AGL (Above Ground Level). Common test conditions consisted of visual meteorological conditions with high visibility in excess of 24 km and rural terrain covered in vegetation. PICAS was installed on the Bell 205 rotorcraft, a medium single engine helicopter with an experimental fly-by-wire control system installed, with the capability to revert to a certified mechanical control system configuration piloted by a safety-pilot. The architecture and system description mirrors that of the NRC's Bell 412 Advanced Systems Research Aircraft (Reference 14).

The Bell 205 (Figure 4(a)) acted as a surrogate RPAS, and flew collision intercepts against multiple fixed-wing intruders (Harvard/Texan, Extra 300, DeHaviland Twin Otter). All aircraft were equipped with a high-precision GPS/INS and a Garmin GDL-90 ADS-B UAT (978 MHz) to record quality truth data. Eager readers could find the flight test technique for conducting collision intercepts, pilot displays, and a seated man-rack described in detail in Reference 15.

a) **Bell 205 – surrogate RPAS**

b) Extra 300

c) **Harvard**

d) Twin Otter

Figure 4. Surrogate RPAS and intruder aircraft

3.2 Truth data collection

Both ownship and intruder ground truth data were separately recorded on each aircraft and were not synchronized in time. Therefore, a tool had to be developed to synchronize the recorded variables to the time when the image was taken by the camera system. The developed Matlab script read the GPS time information for each frame taken from the *.srt* (subtitle) file, and interpolated the intruder and ownship information to that time. This interpolation tool also calculated separate variables of interest, e.g. azimuths to the intruder with respect to track and heading, elevations to the intruder with respect to pitch and flight path, time to closest point of approach (CPA), and distance at the CPA, which are of interest for further data analysis. The results were saved as Matlab structures for each aircraft (ownship and intruder).

To assist in collecting the "pixel truth" data, a tool was developed to help a human record and save true locations of the intruder on the image frame (Figure 5). Pixel truth was considered to be determined by the human after clicking on the location of the intruder aircraft on the image. The tool used calibration parameters for each camera, and read the position of ownship and intruder aircraft at each time when the image was taken. After performing several transformations (rotations around world axis, body axis, camera axis, etc. in the reverse order of that described in Figure 3), the intruder position information was projected into the image plane in the form of pixels with a bounding box of 200 pixels around it. If the user clicked on the box, this part of the image would become zoomed out to make it easier to record pixel truth. Considering that the intruder was not always in the field of view of the particular camera, this tool only presented to the user the image frames with the intruder being in the field of view. This feature helped to reduce time spent on recording pixel truth. The tool also replayed the images in reverse order, making the target initially easy to find, and allowing for visual landmarks from prior detections to assist in the determination of the true pixel location. At the end of each "clicking" session, several variables were recorded in *.mat* format, which could be found in the "pixel truth" folder:

1. Horizontal pixel truth (-1 if the user could not see aircraft on the image);
2. Vertical pixel truth (-1 if the user could not see aircraft on the image);
3. Projected horizontal location of the intruder on the image;
4. Projected vertical location of the intruder on the image.

In order to verify that the human operator correctly determined the pixel truth, a data replay tool was developed (Figure 6). This tool read the recorded user clicked data from *.mat* file, and marked the user clicked intruder position as a "green cross" and back-projected the intruder position as a "red cross" on the image frame, as determined from the INS data and coordinate transformation. By replaying the data in this manner the user was able to visually determine the roll/pitch/yaw offsets resulting from the calibration errors.

Figure 5. Pixel truth recording tool

Figure 6. Data replay tool

4 PERFORMANCE METRICS

4.1 PICAS detection and tracking algorithm

The PICAS algorithm was developed by the NRC under the leadership of Cyrus Minwalla to detect and track General Aviation intruder aircraft. A high-level processing pipeline for PICAS is shown in Figure 7. The detection/tracking algorithm can be summarized as follows:

a) Perform a similarity transform to find the inter-frame homography between the subsequent images, caused by rotation and translation. This was also needed at the tracking stage to apply to previously detected targets.

b) Get the coordinates of the horizon using a WGS-84 ellipsoid model, given ownship position/orientation, as well as known terrain height, and fit them into the line (Reference 8).

c) Detect targets on the image frame:

1. Apply a mask (several options are possible: None, Step, Gaussian, and Skew-normal). This required the horizon line to apply the mask in the appropriate location. If the horizon line was unavailable, the algorithm applied the mask around the vertical middle of the image. Applying different masks was one of the tuning parameters of the algorithm, and was needed to reduce potential false targets from ground clutter below the horizon.
2. Detect hits using the NRC's and built-in detectors. The options were CLF (the NRC's proprietary), FAST, or Shi-Tomasi. These feature detectors were described and compared in Reference 16.

d) For all the detections, compute target metrics such as target signal, background signal, local noise, etc.

e) Update the tracker (tracking was done in image coordinates; the tracker matched targets based on how close and how similar they were to a previous target in the tracker list).

f) Estimate target range; this estimation was based on aircraft altitude, visibility, and the atmospheric extinction coefficient (details can be found in Reference 8).

g) Retain *N* tracks per camera, with the highest closure rates. The number of the tracks maintained directly impacted the processing power and memory resource utilization of the algorithm and was selectable by user (nominally 6).

h) Transform the intruder from image coordinates to geodetic coordinates.

Further details of the algorithm cannot be disclosed in this paper, as it is Intellectual Property of the National Research Council of Canada.

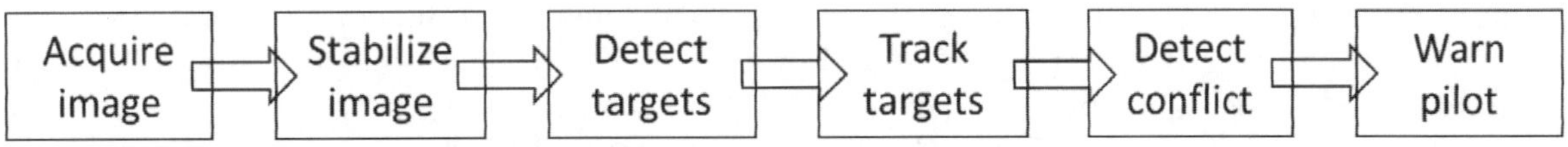

Figure 7. PICAS processing pipeline

4.2 Algorithm performance analysis

This section presents a sample analysis of the performance of the PICAS algorithm on the recorded image set from the Master camera for a single head-on intercept which occurred on December 6, 2016 with Harvard Mark IV acting as intruder aircraft (Figure 8). The Bell 205 ownship was moving with a ground speed of 80 knots, and the intruder was moving with a speed of 140 knots.

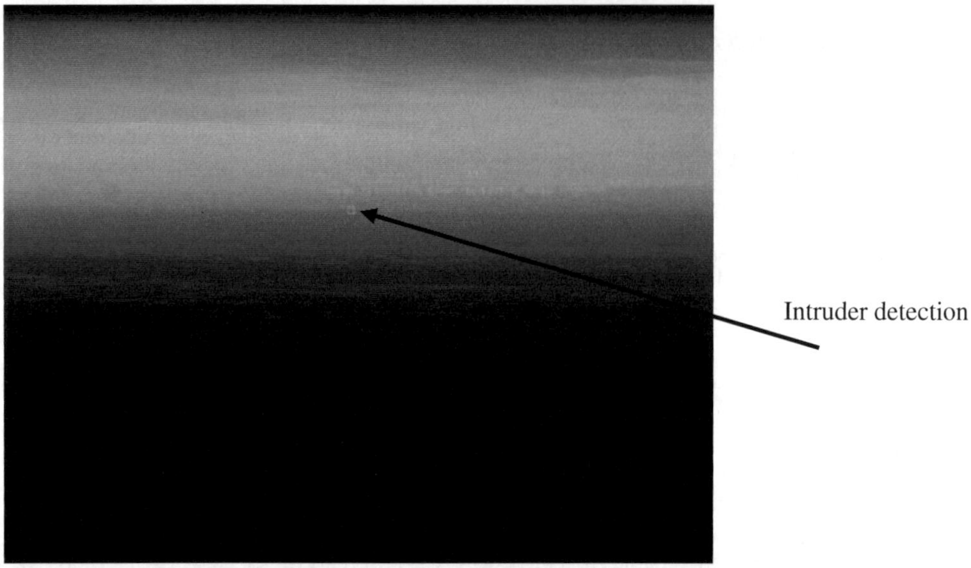

a) Output of the processing software

b) Visualization of the intercept as KML overlay on Google Earth Imagery

Figure 8. Offline data processing

The output of the processing algorithm was intruder detections in pixels as in Figure 8, as well as position estimates of the intruder aircraft (latitude, longitude, altitude). The detection data analysis, on average, revealed 6 pixels deviation from user clicked data. Figure 9 shows the pixel error with the mean calibration error removed.

Overall, 740 images were analyzed by the software, starting from an approximately 7 km range, ending at 1.2 km. To make it easier for the reader, the errors in this analysis are presented as a function of range to the intruder, rather than error at each frame.

Figure 9 shows a surprising result that larger errors in pixel analysis occurred at closer ranges. This may be explained by the fact that the user was likely to consider the "nose" of the intruder aircraft on the image to be the location of the aircraft, while the detection algorithm used the detected feature's "center of gravity". Therefore, as the feature grew in size as the intruder approached there was potential for greater pixel error.

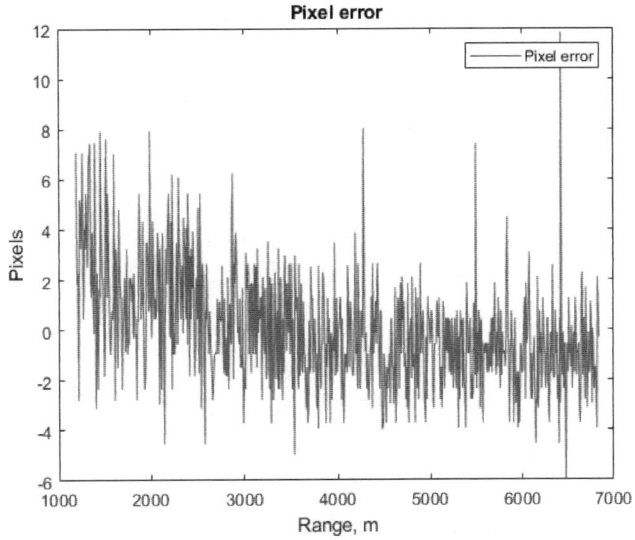

Figure 9. Detection data analysis

The average horizontal position error was 346 m, whereas the average altitude error was 270 ft (shown in Figure 10, means removed).

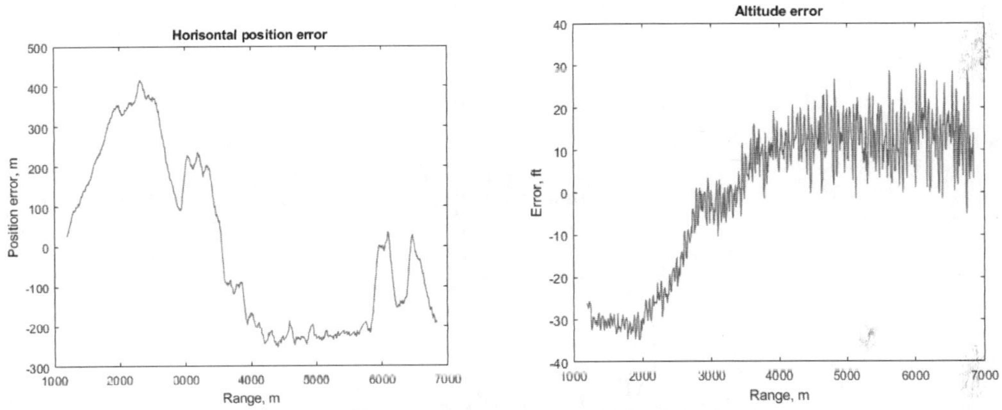

a) **Horizontal position error** **b) Altitude error**
Figure 10. Position error analysis

It is believed that the position error is largely caused by the range estimation error, which is a known challenge for camera-based systems. It can be seen from Figure 11 that the range error (311 m on average) repeats the shape of horizontal position error, as expected, and is a significant contributor to both altitude and horizontal position errors. *Note* that range estimation was tuned for targets which are located further, and was known to be less accurate for close range targets. This is the reason why horizontal position error and altitude error in Figure 10 are much lower at further ranges.

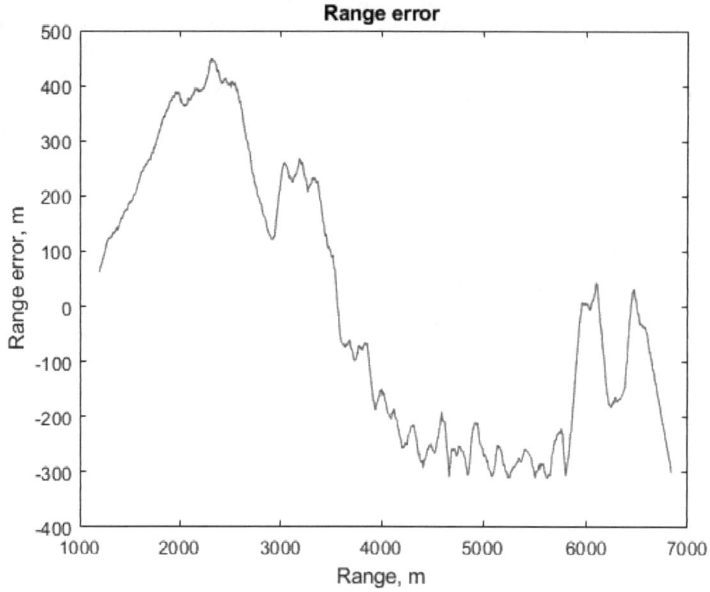

Figure 11. Range error analysis

Mean elevation and mean bearing to the intruder errors were determined from the Pinhole camera model (Eq. 3) to eliminate the impact of range error, and are 0.66 degrees and – 0.72 degrees respectively. Note, that 1 degree error at 7 km range is approximately 120 m (approximately 1/3rd of the average total position error), reinforcing the notion that range error was the largest contributor to position error of PICAS system.

$$\tan(\theta) = \frac{u - x_f}{f},$$

$$\tan(\varphi) = \frac{v - y_f}{f}; \quad , \qquad (3)$$

where θ, φ - bearing and elevation to the intruder, (u, v) -pixel coordinates of the intruder on image plane; f – focal length of camera, (x_f, y_f) - coordinates of the optical centre of the image.

Figure 12 presents the bearing and elevation errors as a function of range. It can be seen that bearing and elevation errors are larger when the ownship is closer to the intruder. This could mean that small target feature-based detection, such as the PICAS algorithm, may result in larger error when the intruder occupies more pixels in the image, making the closer range detection a more suitable problem-space for AI-based algorithms which perform well in a data rich environment. For instance, Figure 12 shows that both elevation and bearing errors start increasing significantly at a range of approximately 4km, which is coincidentally where the target started to occupy more pixels on the image: 40 pixels horizontally and 15 pixels vertically, resulting in potential errors of +/- 0.05 degrees in elevation and +/-0.1 degree in bearing. It can be observed that the largest angular errors were at the range of 1.2 km, where the target occupied about 100 pixels horizontally and 25 pixels vertically, resulting in potential errors of +/-0.07 degrees in elevation and +/-0.3 degrees in bearing.

The fact that the target occupied more pixels horizontally rather than vertically owing to the visible wingspan aligns with the observation that the overall bearing error was larger than elevation. The small target feature detection algorithms used blob based approaches to estimate the center of gravity of the feature being detected. Beyond a certain target size (pixels) the blob-based approach broke down and failed to accurately determine the center of gravity of the detected feature. Another contributor to the bearing and elevation errors could be calibration errors of the camera array. As mentioned in Section 3.2, all the cameras had static calibration errors which were mostly constant throughout all the intercepts. For example, for the "Master" camera used in this example, camera roll offset from the value determined after the calibration was -1.73 degrees, pitch offset was -0.2 degrees, and heading offset was 1.21 degrees, which would result in a static error at all the ranges.

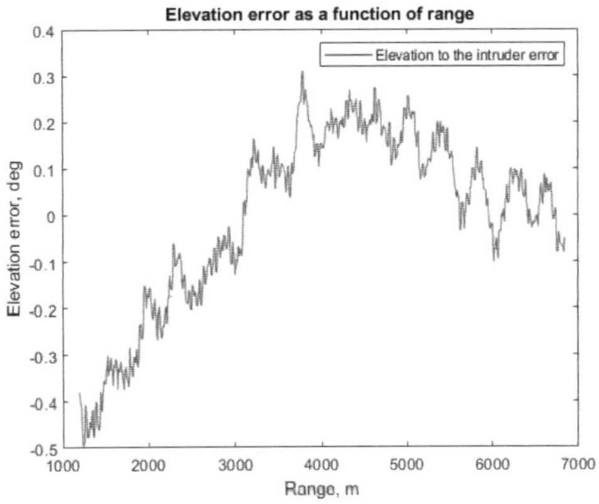

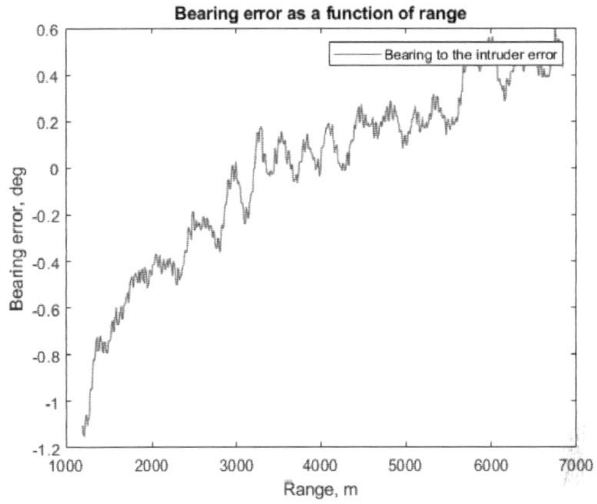

a) Elevation error **b) Bearing to the intruder error**

Figure 12. Elevation and bearing error

CONCLUSION

This paper described the NRC's NMAC imagery database, including: the camera specifications, extrinsic/intrinsic camera parameters, and camera installations on the surrogate unmanned aircraft. It proposed preliminary methods for establishing performance metrics of an electro-optical Detect and Avoid system, which are intended to support performance standards under development. More data analysis will be conducted in future work.

The data analysis showed that feature-based algorithms could detect collision-course General Aviation targets from large distances (7 km). The largest contributor to position error for a single camera DAA system was range estimation error (311 m). Bearing and elevation errors increased as the intruder's range to the camera decreased, which might imply that feature-based detection resulted in larger error when the intruder occupied more pixels on the image, making close range detection and tracking a more suitable problem-space for AI-based algorithms. Camera-based detection was also shown to be sensitive to calibration errors which could impact the accuracy of position estimation.

The NRC's NMAC database and its associated truth data has potential to become a standard set to benchmark electro-optical Detect and Avoid systems and is available freely upon request (contact NRC.NMACDatabase-basededonneesNMAC.CNRC@nrc-cnrc.gc.ca for access).

ACKNOWLEDGEMENTS

Authors gratefully acknowledge Cyrus Minwalla and John White for their contributions to this research project.

REFERENCES

1. A. Zarandy, T. Zsedrovits, Z. Nagy, A. Kiss, A., and T. Roska, "Visual sense and-avoid system for UAVs." *Cellular Nanoscale Net.* 2012.
2. W. Leong , P. Wang , S. Huang , Zh. Ma, H. Yang, J. Sun, Y. Zhou, M. Redhwan, A. Hamid, S. Srigrarom , and R. Teo, "Vision-Based Sense and Avoid with Monocular Vision and Real-Time Object Detection for UAVs." *2021 International Conference on Unmanned Aircraft Systems (ICUAS).* Athens, Greece, June 15-18, 2021.
3. Y. Zhang, "Aircraft Detection and T aft Detection and Tracking Using UAV-Mounted Vision - Mounted Vision System", *Master of Science in Electrical and Computer Engineering December,* 2015, Embry-Riddle Aeronautical University Daytona Beach, Florida
4. FlightGear Flight Simulator, Available at: https://sourceforge.net/projects/flightgear/files/release-2020.3/; retrieved: March 2022.
5. X-Plane, Available at: https://www.x-plane.com/manuals/desktop/; retrieved: March 2022.
6. J. Keillor, K. Ellis, G. Craig, D. Rozovski, and R. Erdos, "Studying Collision Avoidance by Nearly Colliding: A Flight Test Evaluation." *Proc. Human Factors and Ergonomics Annual Meeting*, 2011, pp. 41-45.
7. I. Borshchova and K. Ellis, "2021 NMAC database", Available at: https://nrc-digital-repository.canada.ca/eng/view/object/?id=2a82aa4a-4e08-41d5-8938-8ebbb0509b1f; retrieved: March 2022.
8. C. Minwalla, P. Thomas, R. Hornsey, K. Ellis, and S. Jennings, S., "Range performance evaluation from the flight tests of a passive electro-optical aircraft detection sensor for unmanned aircraft systems." *Journal of Unmanned Vehicle Systems*, Vol. 4, No. 2, Feb. 2016, pp. 96-114.
9. Allied Vision 2022, Available at https://www.alliedvision.com/en/products/camera-series/prosilica-gt/?gclid=CjwKCAiApfeQBhAUEiwA7K_UH023Q0u8TxOZ3zqlDId6DAwaBWIR9aM0ffPqJPQ3eEHKtkFBs_P2ihoCIewQAvD_BwE; retrieved: March 2022.
10. J-Y. Bouguet, "Camera Calibration Toolbox for MATLAB."Available at: https://www.vision.caltech.edu/bouguetj/calib_doc/; retrieved: March 2022.
11. R. Hartley, and A. Zisserman, A., "Multiple View Geometry in Computer Vision." *Cambridge University Press*, 1st ed., 2003.
12. Z. Zhang, Z., "Flexible Camera Calibration by Viewing a Plane from Unknown Orientations." *Proc. IEEE International Conference on Computer Vision*, 1999, pp. 666-673.
13. A. Zisserman, "Matlab functions for multiple view geometry." Available at: https://www.robots.ox.ac.uk/~vgg/hzbook/code/, retrieved: March 2022.
14. A. Gubbels, and S. Carignan, "The NRC Bell 412 Advanced Systems Research Aircraft - A New Facility for Airborne Simulation." *Canadian Aeronautics and Space Journal*, Vol. 46, No. 2, 2000, pp. 106-115.
15. K. Ellis, I. Borshchova, S. Jennings, and C. Paleske, "A comparison of two novel approaches for conducting detect and avoid flight test." *Journal of Unmanned Vehicle Systems,* Vol. 9, pp. 149–165, dx.doi.org/10.1139/juvs-2021-0005.
16. D. Tulpan, N. Belacel, F. Famili, and K. Ellis, "Experimental evaluation of four feature detection methods for close range and distant airborne targets for Unmanned Aircraft Systems applications." *2014 International Conference on Unmanned Aircraft Systems (ICUAS)*, May 27-30, 2014, Orlando, FL, USA.

AUTHOR INDEX